D1486743

Technology Transfer
and Human Factors

Other Titles of Interest from Lexington Books:

Investing in Developing Countries
A Guide for Executives
Thomas L. Brewer, Georgetown University;
Kenneth David, Michigan State University; and
Linda Y.C. Lim, University of Michigan
Foreword by Gerald T. West
ISBN 0-669-12770-1   272 pages

Multinationals, Technology, and Industrialization
Implications and Impact in Third World Countries
Katherin Marton, Fordham University
ISBN 0-669-13209-8   320 pages

Multinational Corporate Strategy
Planning for World Markets
James C. Leontiades
Foreword by H. Igor Ansoff
ISBN 0-669-07381-4   256 pages

International Joint Ventures
An Economic Analysis of U.S.–Foreign Business Partnerships
Karen J. Hladik, Exxon Corporation
ISBN 0-669-09907-4   144 pages

Country Risk
Assessment and Monitoring
Thomas E. Krayenbuehl, Union Bank of Switzerland
ISBN 0-669-10958-4   192 pages

Technology Transfer and Human Factors

Charles T. Stewart, Jr.
The George Washington University

Yasumitsu Nihei
Keio University

Lexington Books
D.C. Heath and Company/Lexington, Massachusetts/Toronto

Library of Congress Cataloging-in-Publication Data

Stewart, Charles T., 1922-
  Technology transfer and human factors.

  Includes bibliographies and index.
  1. Technology transfer—Developing countries. 2. Technology transfer—United States.
3. Technology transfer—Japan. I. Nihei, Yasumitsu, 1933–   . II. Title.
T174.3.S74   1987   338.9'26   86-45751
ISBN 0-669-14251-4 (alk. paper)

Published simultaneously in Canada
Printed in the United States of America
Casebound International Standard Book Number: 0-669-14251-4
Library of Congress Catalog Card Number: 86-45751

The paper used in this publication meets the minimum requirements of American National
Standard for Information Sciences—Permanence of Paper for Printed Library Materials, ANSI
Z39.48–1984. ∞ ™

87 88 89 90 8 7 6 5 4 3 2 1

Contents

Tables

Preface and Acknowledgments

The United States–Japan Advisory Commission, established in May 1983 by the governments of the two countries to advise on the conduct of U.S.–Japan relations, submitted a report in September 1984. One of its conclusions was that "[e]conomic development and political stability in the developing world are matters of vital concern to the United States and Japan for economic, political, and humanitarian reasons" (United States–Japan Advisory Commission [U.S.–J.A.C.] Report 1984, 90). Development assistance is a major source of finance for long-term infrastructure projects essential to the development process. However, foreign direct investment is also essential to increase the supply of available capital, stimulate economic diversification and technology transfer, and enhance employment opportunities. "Consultation between the United States and Japan in regard to aid programming has been close and increasingly effective" (U.S.–J.A.C. Report 1984, 94). It recommended that "the two countries should continue to seek opportunities to enhance the impact of their individual bilateral-aid efforts by identifying areas in which they can coordinate or dovetail projects, taking advantage of experience and particular expertise. . . . Attention should also be focused on cooperative science and technology programs in appropriate fields related to national development. . . . The United States and Japan should consider scientific cooperation to provide technology suited to the capabilities of developing countries" (U.S.–J.A.C. Report 1984, 94, 100).

The research project described in this book was conducted in the spirit of these findings and recommendations. It is a comparative study of technology transfer by Japan and by the United States to less-developed nations. It focuses on the human resource dimension of technology transfer, using case studies of Indonesia and Thailand. It is our expectation that the findings will be widely applicable to other Third World nations. The objective of the research is to recommend ways of improving the process and increasing the rate of technology transfer and, as feasible, to suggest ways in which Japan and the United States may cooperate toward this end.

The research was conducted by Yasumitsu Nihei, professor at the Institute of Management and Labour Studies, Keio University, Tokyo, and by Charles T. Stewart, Jr., professor of economics at The George Washington University, Washington, D.C., through the Institute for Sino–Soviet Studies of The George Washington University. The research was sponsored by the U.S.–Japan Foundation and the Japan–United States Friendship Commission.

The book is based on interviews conducted by Professors Nihei and Stewart on two trips to Indonesia and Thailand, as well as on interviews done in Tokyo and Washington. Also incorporated were a review of reports by the four governments concerned and by some nongovernment organizations involved in technology transfer; a survey of relevant scholarly work; questionnaires to U.S. and Japanese firms; and the suggestions of numerous individuals who participated in symposia in Tokyo and Washington in January 1984 and conferences in May 1985. This book is incomplete in that a thorough study of the entire subject is beyond the limited resources of the researchers.

The book is organized into four general parts. Chapter 1 outlines the concepts and definitions that guided our study, and the boundaries we selected on the scope of our research. It describes the incentives and deterrents to technology transfer, and lists the major policies in less-developed countries (LDCs) that can affect the amount and type of technology transfer. Chapters 2 and 3 describe the environment in Indonesia and in Thailand for technology transfer. They discuss briefly the economy, resources, and labor force; the major problems, the plans and objectives of the national governments; and public policies affecting technology transfer. Chapters 4 through 6 describe the contributions that have been and are being made by U.S. and by Japanese organizations to technology transfer to Indonesia and to Thailand (these include nonprofit organizations and business firms as well as government organizations) and discuss the differences between the U.S. and Japanese contributions. Chapter 7 presents our conclusions, recommendations, and suggestions for increasing the level and improving the process of technology transfer by U.S. and Japanese organizations and for greater cooperation between these two countries to that effect.

Our work was done in collaboration with the Human Resource Institute of Thammasat University and its director, Dr. Chira Hongladarom, who were responsible for obtaining information on U.S. firms in Thailand. We also worked with the Japanese Chamber of Commerce in Bangkok and the Indonesia–Japan Enterprises Association in Jakarta, who were responsible for collecting information on Japanese firms in Thailand and Indonesia respectively, with the cooperation of the local offices of the Japan External Trade Organization and the support of the Ministry of Foreign Affairs of Japan.

We wish to acknowledge the financial support of the U.S.–Japan Foundation and the Japan–United States Friendship Commission, and the institutional support of the Institute for Sino–Soviet Studies of The George Washington University, through which this research was conducted. We thank especially

Gaston Sigur, director of the institute, and William Johnson, acting director while Professor Sigur was on leave, for their encouragement and leadership. We are indebted to the participants in symposia and conferences in Tokyo and Washington who reviewed earlier drafts; to Yasuhiko Torii, who organized the Tokyo meeting; and to numerous individuals whom we interviewed in Indonesia and Thailand. We do not list these individuals for fear of omitting some we should include, or including others who may prefer to remain nameless. We exonerate all of the above from any responsibility for the final product. We thank Suzanne Stephenson and Dorothy Wedge for threading bureaucratic mazes and arranging conferences; Jin-Hsia Lee for patient and valuable assistance in research; and Leila Sanders and Sheila Murphy for guiding the manuscript from early drafts through its final version.

Reference

United States–Japan Advisory Commission, 1984. *Challenges and Opportunities in United States–Japan Relations*. Report submitted to the president of the United States and the prime minister of Japan, September.

Abbreviations

AID	Agency for International Development (United States)
AIT	Asian Institute of Technology
APCAC	Asia–Pacific Council of American Chambers of Commerce
ASEAN	Association of Southeast Asian Nations
ATI	Appropriate Technology International
BKPM	Investment Coordinating Board (Indonesia)
BOI	Board of Investment (Thailand)
DOD	Department of Defense (United States)
EEC	European Economic Community
ESCAP	Economic and Social Commission for Asia and the Pacific
FMS	Foreign Military Sales (United States)
FMME	Fund for Multinational Management Education
GDP	gross domestic product
GNP	gross national product
GOI	Government of Indonesia
IESC	International Executive Service Corps
IMAT	International Military Education and Training (United States)
IRRI	International Rice Research Institute
JETRO	Japan External Trade Organization
JICA	Japan International Cooperation Agency
KDI	Korean Development Institute (South Korea)
LDC	less-developed country
LNG	liquified natural gas
LPEM	Lembaga Pendidikan Ekonomi Manajemen (Institute for Management Economics Education)
LPPM	Institute of Management Development and Education (Lembaga Pendidikan Kan Pembinaan Manajemen—Indonesia)
MDC	more-developed country
MNE	multinational enterprise
NIC	newly industrializing country
OPEC	Organization of Petroleum Exporting Countries
OPIC	Overseas Private Investment Corporation
PRC	People's Republic of China
PVO	private and voluntary organization
RTG	Royal Thai Government
TAICH	Technical Assistance Information Clearing House
TDP	Trade Development Program
VAT	value-added tax
VITA	Volunteers in Technical Assistance

1

The Process of Technology Transfer
via Human Resource Development

This chapter outlines the conceptual framework within which this study was conducted. What do we mean by technology transfer? What is the human dimension of technology transfer? How can we measure or indicate the transfer of technology via human resource development? What are the processes whereby technology is transferred? What are the roles of different types of institutions in contributing toward technology transfer? What factors determine the amount and type of technology transferred? What are the constraints on technology transfer, and how may they be relaxed?

In later chapters we will examine the types of constraints found in Indonesia and Thailand. We will consider the potential contribution of Japanese and U.S. organizations toward relaxing these constraints, as well as the factors promoting technology transfer to these countries, and how these factors may be enhanced.

Technology and Its Transfer

Both *technology* and its *transfer* have been given such a wide variety of meanings that we need to specify first what we mean by these terms. *Technology* refers to new and better ways of achieving economic ends that contribute to economic development and growth. We deliberately exclude technologies unrelated to or only remotely related to development and growth. A further limitation concerns the types of technology to be considered.

> The simplest version views a technology as involving only changes in artifacts. A more sophisticated approach adds to the physical objects, labor and managerial skills. . . . A third approach views technology as a "socio-technological" phenomenon; that is, besides involving material and artifact improvements, technology is considered to incorporate a cultural, social, and psychological process as well. (Murphy, in Spencer & Woroniak 1967, 6–7).

We shall focus on the second dimension, the human resource dimension, which has been neglected relative to the physical dimension in many LDCs. It has thus become the major constraint in transferring technology. The third, or sociocultural dimension, is also important; it has been neglected because it is less tangible than the others, more country and group-specific, and less easily and more slowly subject to modification. We were (or became) aware of this dimension in our work, but could not, for lack of time and expertise, thoroughly examine its role in facilitating or inhibiting technology transfer. We shall refer to this third dimension from time to time, but will not consider it systematically.

Technology transfer is the "utilization of an existing technique in an instance where it has not previously been used" (Gruber & Marquis 1969, 255). It is to be distinguished from diffusion of an already-established technology within a country. In practice, however, transfer and diffusion form a continuum, with no clear dividing line.

> The transfer concept must include an aspect of "absorption" of the technology on the part of the people of the recipient country. Therefore, international transfer of technology has not taken place until the technology under consideration has been assimilated to some degree by the recipients in the host country. . . . Assimilation is the recipient's capacity to utilize the technology on his own. . . . National diffusion, within the recipient nation, occurs when nationals leave the original transfer channel (i.e., the MNC) for other jobs within the national economy in which they utilize the acquired technology, or when still other nationals acquire the technology by whatever means. (Konz 1980, 21, 22, 25)

The key distinction between technology transfer and technology diffusion is that technology transfer concentrates on the supply side—on the willingness and ability of the supplier to transfer—and assumes that demand is not a constraint. Diffusion on the other hand takes technology supply for granted, and examines the spread of demand over space and time. There is a corresponding difference in the type of technology considered. Standard, open technology is freely available to those who have the wish and capacity to absorb it; its supply is unlimited, and demand alone determines the rate of diffusion. On the other hand, some technologies offer prospects of monopoly rents (patents, trademarks, or exclusive know-how). These are technologies with development costs that have not yet been fully recovered; those that are still the property of a limited number of firms; or those whose exercise still depends on the know-how of a limited number of technical and managerial personnel. This type of technology has value to its possessor, and is not going to be transferred without substantial cost and appropriate payment—if it is transferred at all.

Diffusion is also used to mean the spread of the *use* of a technology rather than of the technology itself. Increasing the number of computers in a country does not involve transfer of IBM technology, but only an expanding use of it. The only technology transferred is the use of computers, not their design or manufacture. IBM is only too willing to provide this technology, as it expands the market for IBM products. Likewise, the spread of the use of hybrid seed in farming does not imply transfer of the knowledge for developing or producing hybrid seed, but only for using it effectively. As long as a firm reaps economic rents from its monopoly in a product or process (whether this be a patent, a trade secret, or simply early experience placing the firm higher on the learning curve than its competitors), the firm will be unwilling to transfer the technology at all, or only at a price sufficient to compensate it for the loss of economic rents. The firm only too willing to sell its patented pharmaceuticals or its hybrid seed is unwilling to allow others to produce the drug in competition with it, or to reveal to other firms the genetic components of its hybrid seed. In sum, diffusion is that aspect of the spread of technology that involves no economic rents, whereas transfer is that aspect that does.

In terms of this book, technology transfer means direct foreign investment and turnover of local employees who have mastered the technology to implement it in local organizations. Technology transfer also covers the licensing of domestic firms by foreign firms to produce products or to use processes for which the foreign firm receives some compensation and provides training and technical assistance. Turnkey plants also involve technology transfer insofar as local workers learn to build such plants, and through turnover exercise this investment capability elsewhere in the host nation (and, in a more limited sense, insofar as local managers and workers need to be trained to operate them). Technology transferred from one domestic firm to another, regardless of how it was acquired (licensing, turnover of multinational enterprise [MNE] employees, etc.), is internal, and not transferred from foreign organizations. Technology acquisition not accompanied by (or not requiring) specific contributions from the possessors of technology (for example, training, technical assistance, or licensing), is technology diffusion, not transfer.

Dahlman and Westphal (1983, 7), among others, distinguish between three levels of technology transfer: (1) the capability required to operate a technology, for example, to run and maintain a plant; (2) investment capability—that required to create new productive capacity; and (3) innovation capability—the ability to modify and improve methods and products. These require different types and levels of skill, and different supporting institutions. Operational capability is short-term; even investment capability may not last, because products and production techniques become obsolete. Only when all three capabilities have been transferred has the receiving nation acquired a permanent mastery of the technology. Operational capability can be learned on the job. Investment capability requires formal training; on-the-job training is protracted, and what is learned needs to be adapted, not just replicated. Innovative capability cannot be acquired from work experience alone. Some innovations presuppose a technical education. Many are not highly technical, but call upon imagination and a mental set always seeking better ways.

There are precursor and predisposing conditions for technology transfer, and these are often lumped together with technology transfer. One precursor condition is general knowledge, disseminated through education to a growing portion of the population. General knowledge prepares individuals to acquire technology-specific know-how required for new production processes or products. It represents absorptive capacity. There is no sharp dividing line in many cases between general knowledge (as a prerequisite) and technology-specific know-how, or between education and training.

Another kind of precondition is sometimes included in technology transfer when the concept is stretched to its limits: the propensity to adopt new products, which creates markets for them, and the propensity to adopt new techniques, without which absorptive capacity remains underutilized and technology transfer does not take place. This precondition is highly correlated with general knowledge, for information is one ingredient in willingness to accept change. Attitudes and values embedded in the culture and religion, and in the institutional structure of society, may inhibit change. The accommodation or modification of such attitudes may be a precondition for accelerated technological transfer. (See Niehoff & Anderson 1964; Novack & Lekachman, eds. 1964, part II).

Economic growth and development might be described as a product of acceptance of change, of absorptive capacity, and of supply of new technology. If any of these factors is zero, the product, technology transfer, is zero. Thus concern with technology transfer does encompass issues of absorptive capacity and acceptance of change. But this book must be limited in scope, mainly to technology transfer in the narrow sense, encompassing technology-specific human resource development. Some attention is devoted to activities affecting absorptive capacity insofar as they impinge on technology transfer, but no effort is made to study absorptive capacity in depth. Neither will we probe further, for example, into the values, attitudes, and institutions that resist or facilitate technology transfer.

Types of Human Resource Development Complementary to Technology Transfer

The two broad types of human resource development needed to effect technology transfer correspond roughly to Gary Becker's distinction between general and specific training. General education and training are prerequisites for effective technology-specific training. Workers may have to be literate and numerate; technical and professional employees may need an engineering background;

managerial employees may need a substantial general education plus exposure to a variety of management-specific topics. This kind of preparatory education we call absorptive capacity. It sets limits to how much technology and what kind of technology can be transferred by determining the feasibility and costs of technology-specific training.

Absorptive capacity has many dimensions, which, however, may be aggregated into: (1) the supply of workers with the general educational background appropriate for hiring and technology-specific training by MNEs at the technical and managerial level; and (2) the state of the legal-social-economic infrastructure of the nation. This second kind of absorptive capacity is institutional rather than individual, but its existence is predicated on a supply of appropriately educated and trained individuals. It also sets limits as to how much and what kind of technology can be transferred at acceptable risk and cost. "No developing country can make adequate use of technology flows unless it has an appropriate receiving system composed of a scientific and technical infrastructure in the public, private, educational, and corporate sectors. It is of no value to develop a technical training system or R&D institutes when companies are not interested in using them" (Fund for Multinational Management Education 1978, vol. 2, 55).

Absorptive Capacity

"The real trouble is not merely a shortage of specific skills, but a more general and pervasive lack of skills and abilities to digest, absorb and diffuse the modern technology" (Myint, in Spencer & Woroniak 1967, 156).

Absorptive capacity is created primarily by the host country. It is a fundamental determinant of how much investment and firm, industry, and technology-specific training MNEs find it worthwhile to undertake. This in turn influences what kinds of products and technologies MNEs are prepared to introduce into an LDC, and how much and what kind of investment a host country can attract. Training conducted by firms must build on an educational foundation provided in and mainly by the host country, if training is to include higher managerial and technical personnel from the home country. MNEs are not normally in the business of providing the general education without which the technology-specific training is impossible or inordinately costly.

Educational System. The relevant educational background will include engineering specialties, business administration, and related fields. Other technical fields are necessary, depending on the human resource requirements of the technology to be transferred (for example, geologists for oil exploration). If MNEs cannot hire adequately educated people, then either they must rely heavily on expatriates (which is expensive and may run into restrictions on the use of expatriates), or they must undertake general training of new hires to prepare them to absorb technology- and industry-specific training. In either case, start-up and operating costs become high—at some point prohibitively high. Absorptive capacity can be estimated with reference to specific industries, keeping in mind that some educational backgrounds (business administration, for instance) prepare graduates for training in a wide range of industries, whereas others (geology) do not. T extiles, for example, has limited requirements for managerial and technical skills compared to petrochemicals; hence a country may have a large absorptive capacity for textiles, and none for petrochemicals. Another way of expressing this difference is to say that investment by MNEs in textiles involves little technology transfer in many LDCs (none in some), whereas petrochemicals typically involves very substantial technology transfer potential.

Because absorptive capacity in human resources is created mainly by a country's educational system, increasing the number and/or quality of graduates is the main way of relaxing human resource constraints on technology transfer. If universities are to graduate adequately prepared engineers and managers, clearly they must draw upon a sufficient supply of students who have benefited from a satisfactory primary and secondary education—the foundation on which universities must build.

Infrastructure. The second category of absorptive capacity, the legal-social-economic infrastructure, provides services rather than human resources to business, and specifically to MNEs. These services include: (1) the supply of and cost of transportation, communications, utilities, financial services, as well as facilities (such as materials testing laboratories), that are sector- or industry-specific; and (2) the services of government in regulating economic activities, providing information, setting standards, administering the legal-social environment of business, and operating the fiscal machinery that both helps regulate the economy and provides revenues for its services. The provision of education is also a major function of government. Inadequate supply of business services, such as transportation, may compel the MNE to supply its own at higher cost or to suffer the consequences of inadequate supplies, delays, and interruptions in production and delivery. In either case the cost of doing business goes up. Shortcomings in government administration and regulation may be reflected directly in business costs, or more indirectly as a climate of uncertainty surrounding the activities of MNEs. Such a climate is always a deterrent to investment and hence to technology transfer.

Absorptive capacity in terms of business services and government administration is largely created by the educational system. Both private nonprofits and public foreign aid donors have contributed significantly through education and training assistance, particularly in the area of public administration. Foreign firms engaged in the construction and equipping of many economic infrastructure projects (transportation, communications, and utilities), prepare indigenous personnel to operate and maintain the facilities. Considerable technology transfer is involved in this process.

Business Investment, Training, and Technology Transfer

We know from a generation of research on economic growth in the United States and other advanced countries that most technological progress is associated with investment in plant and equipment. Not much progress can be attributed to improvements in technical and managerial knowledge unaccompanied by changes in physical capital. We also know, from the research of Fellner (Fellner 1970) and others, that changes in products and production processes almost invariably require new skills and large increases in managerial inputs. Technology transfer from U.S. and Japanese organizations to Thailand and Indonesia and other LDCs then consists mainly of investment in capacity to produce products new to these countries and in productive processes more technologically advanced than those used in these countries. An essential and important part of this investment consists of training employees in the technologies introduced. The number trained, and of course the sophistication of the training, are indicators of the magnitude of technology transfer.

Technology transfer requires the willingness to transfer on the part of the U.S. or Japanese organization, implemented via decisions to invest, the choice of products and productive processes in which to invest, and the complementary training of citizens of the recipient nation to understand and use the technology to be transferred. Very little new technology (new to the recipient country) is so automatic that it can be transferred without complementary skills and human resources to use it and to maintain it. Implantation of a technology new to a country exclusively through the use of expatriates involves no transfer of technology, only the choice of site for its employment.

> Much of the technology that flows across national boundaries is not effectively transferred, on two counts: Control over it is not passed from one set of institutional actors to another, and the capacity to take further steps in elaborating technological modifications or improvements is likewise not vested in new hands, in particular indigenous Third World ones. (Goulet, in Sagafi-nejad 1981, 321)

The kinds of training that are complementary—if not essential—to technology transfer, and by implication the kinds of absorptive capacities that are needed, depend on the level of technology transfer, whether operational or involving investment and innovation capability.

What are the skills essential and sufficient for technology transfer? We are not referring to all skills that may be used in a production process, including those that are commonplace in the receiving nation, or for which the receiving nation has substantial training capabilities. The skills essential and sufficient for technology transfer are those required to form new firms using the new technology, or to modernize existing domestic firms. These are primarily managerial and professional skills. If they are available, the craft, operative, and service skills can be generated; in their absence, even an abundance of production worker skills will not induce technology transfer. This is only a rough-and-ready distinction between the skills essential for technology transfer (created by the firms transferring the technology), and other skills (although needed for the production process, can be supplied by the educational system).

We avoided a survey of all training activities conducted or financed by Japanese and U.S. organizations, including those that merely supplement or contribute to the efforts of the LDC governments and other domestic training and educational institutions. A more abundant supply of such skills may facilitate diffusion of technologies already transferred, but this does not constitute technology transfer. Critical technical skills in acutely short supply may be found in Indonesia and Thailand; we should perhaps include this possibility in our analysis. However, our initial approach was to define relevant training as managerial and professional, and to exclude production worker training, except when reports from MNEs indicated that particular skills should be included.

> The lack of general management resources in the local environment is a major constraint to technology acquisition and utilization. The number of individuals with engineering and production skills far outnumber those with general planning, controlling, and organizing skills. Without these management skills, it is impossible to organize the firm to utilize and exploit technology improvements over the long run. (Wallender 1979, 47)

Technology Transfer Agents

Technology transfer may be accomplished by a diversity of agents, and through a variety of mechanisms. Agents may be organizations in the receiving country, such as business firms, government agencies, or universities. Agents may also be foreign organizations, such as firms, foundations, and similar nonprofit organizations, including universities; international organizations such as the World Bank, and the Asian Development Bank; and government organizations, such as the United States' Agency for International Development (AID) and Japan International Cooperation Agency (JICA). In the former (domestic) case, the initiative lies on the demand side; in the latter (foreign) case, on the technology supply side.

It is suggested that at the earliest stages of development the initiative lies predominantly on the side of the suppliers of technology; LDCs at that stage may indeed request technology transfer, but are not in a position to conduct it themselves. As countries advance and develop, they increase their ability to absorb a wider range of technology, and eventually the initiative passes largely to the firms and government in the LDC, now an NIC (newly industrializing country) if not already an MDC (more-developed country). (Even in the most advanced countries some technology transfer is done on the initiative of the suppliers.)

Our concern is primarily with agents on the supply side and in particular with U.S. and Japanese organizations. The distinction between three kinds of organizations—business firms, government agencies, and private nonprofit organizations—is important because they differ greatly in their goals and in the opportunities and incentives to which they respond. They also differ in the nature of their contribution to economic growth and technology transfer in LDCs. This is particularly true of the U.S. organizations.

The main objective of some nonprofit organizations is to increase the absorptive capacity of LDCs by contributing to education, research, and supporting activities. The main function of others is technology transfer via training and technical assistance. Other nonprofits, whose goals may be community development, relief, or assistance to specific groups, may incidentally contribute to technology transfer via human resource development, although that is not their aim.

MNEs are the main sources of technology transfer (including technology-specific training), but technology transfer is not their objective. They seek markets, supplies, and profits. This is true even of those firms—engineering consulting firms, or contractors—whose business is building plants and training local workers to operate them, or to provide management and technical assistance services. Profit is their primary goal. Firms whose technology is proprietary are reluctant if not averse to transferring technology without adequate return.

JICA and AID certainly contribute handsomely to the economic development of LDCs, but the claim that that is their primary aim is questionable. The term *economic development* is sufficiently elastic to be defined by various agencies to fit their agenda of the moment. The contribution of JICA and AID to economic development is largely a by-product of pursuit of other goals.

Other U.S. and Japanese government agencies encourage business investment in LDCs and stimulate trade, incidentally providing the training associated with the technology transfer, but again, that is not their primary intent.

These comments are intended as a warning. The use of total expenditures (or other indicators of aggregate activity) of organizations with diverse goals is sometimes an inadequate or less meaningful indicator of technology transfer and complementary human resource development, particularly when technology transfer is not a primary concern of the organization. Such indicators will be mentioned, for lack of suitable disaggregation, but should be considered with reservation.

Technology Transfer Mechanisms

The mechanisms for technology transfer are as varied as the agents. This variation depends in part upon the type of technology. Is it proprietary or not? Is it highly sophisticated or not? It depends in part upon the agent undertaking technology transfer. Is it a firm, a government agency, nonprofit, or university? The firm has the capability of using the technology at a new site for productive purposes, whereas the government agency or foundation or university may not be producers. Hence, the firm has options for technology transfer (such as direct investment in a subsidiary) that are not open to other agents. Finally, it depends on the capability of the recipient of the technology to be transferred (a capability which also influences the choice of technology to be transferred). A capable recipient has all options; a recipient that lacks the capability is restricted.

Among possible mechanisms are direct investment by an MNE, which provides the training to provide the skills needed to implement the technology. This may be a wholly owned subsidiary, or a joint venture with a local firm. There is a large possible variation in responsibility between foreign and local partners, depending largely on the capabilities of the latter and partly on local legal requirements. This is the dominant process of technology transfer in the early stages of development. (On stages in the transfer of technology see Kim 1980). It is the most effective means of introducing both hardware and software technologies and developing local human resources to use them.

Although most of the training provided by the MNEs is for their own employees and employees of joint ventures, some training is also provided for local suppliers. Training for local suppliers accomplishes technology transfer to domestic firms without the intervening step of turnover of trained and experienced personnel.

The product or the productive process may be proprietary, either in the legal sense of ownership of patents or trademarks, or in the functional sense that cost-effective production of an item of adequate quality is only possible with the assistance of a firm that is already doing this. (In many cases this is the developer of the product and/or process). A firm in an LDC may purchase the right to produce the product or use the productive process—either outright or through the payment of fees and royalties. The amount of training or technical assistance required can vary substantially, but presumably some of both will be required in all cases.

Most licensing is by MNEs to affiliated firms (Root, in Sagafi-nejad 1981, 122), so direct foreign investment and licensing of firms in LDCs are closely related more often than not. As LDCs improve their absorptive capacity, more technology transfer is via licensing, and more licensing is to independent firms rather than to affiliated firms. Franchises and trademarks are primarily transfers of reputation or goodwill rather than technology, although they include some transfer of managerial technology required for quality control, and generally involve some training and technical assistance.

Further along the development spectrum (keeping in mind that direct investments retain a role at all stages of development where allowed), the local producing organization may simply import productive equipment and place it in operation. This of course requires prior knowledge of the equipment and process, or the educational background to permit self-training in its use. Foreign business firms also provide some training for buyers, for instance, on the use and maintenance of equipment. Much of this training is provided without the intermediation of investment and production in the LDC, via technical assistance from the manufacturer, or from a local distributor who in turn has been trained by the manufacturer. But such training is of use only to individuals with an appropriate technical background. Productive facilities that can be purchased on the international market reflect technology that is open and available to all with the knowledge to understand it. The imported technology may be a product rather than productive equipment; the importing organization may be able to create the productive facilities for its reproduction. The ability to buy a technology and implement it effectively (or to reproduce it through "reverse engineering") includes at least some investment capability as well as operational capability. Whether or not it requires some innovation capability depends on the technology: whether it is static or evolving, and whether it is suited to the LDC or must be first modified.

Turnkey investments are those in which an organization constructs a productive facility, provides necessary training, and then turns over operation entirely to a local firm or other organization. These investments require greater local capabilities to implement new technology, or are limited to technologies that impose lower demands on local capabilities than those that can be provided via direct investment in which the foreign MNE retains a managerial role. Investments funded by international organizations and government agencies are basically of the turnkey nature. In terms of types of technology transfer, turnkey investments may be limited to operational capability, without the investment and innovation capabilities that are typically associated with direct investment.

A producer in an LDC may go beyond purchasing rights to produce a product or to use a particular technology. The producer may buy the foreign producer outright, acquiring the technology, technical personnel, and productive facilities without any transfer in a geographical sense. This transfer may then occur, but as a process internal to the LDC organization.

In concentrating on technology transfer via human resource development by Japanese and U.S. organizations, an implicit decision has been made to concentrate on specific agents of technology transfer—and on specific processes—that are believed to be of greatest importance to the recipient countries studied, Thailand and Indonesia. The principal agents are business firms. The principal mechanism is direct investment and accompanying training and other human resource development activities needed to accomplish the technology transfer required (whether to employ particular productive processes and/or to produce particular goods and services). Other agents certainly are not excluded; in fact, they will be surveyed. We recognize that other mechanisms exist, but we will not study them. Indonesian and Thai organizations (mostly governments and business firms) will not be studied as technology transfer agencies, but only as promoters and constraints on technology transfer by U.S. and Japanese organizations.

The Process of Technology Transfer by Foreign Direct Investors

> Multinational corporations . . . are the dominant institutions transferring industrial technology across national borders. They do so through the sale of their products and services, by training the end users, by investing in local production, by training local employees and technicians, by offering technical assistance to both local customers and suppliers, and by introducing the methodology of integrated research, development, and engineering (Singh 1983, 43).

Technology transfer via human resource development by foreign business firms takes place in three steps. The first is the recruitment and training of local workers in the skills (primarily managerial and professional) required to master and implement the technology used by the firm, but including other types of workers whenever essential. The second step is the advancement of the workers thus trained to positions of greater responsibility as they gain experience, gradually replacing the expatriates who are initially needed both to train them and to perform managerial and professional functions in the firm. This process of replacing expatriates over time involves progress from operational capability to investment and innovation capability. However, this is a protracted process.

The third step, which can occur at every level of technology transfer capability, is the turnover of trained and experienced managerial and technical personnel insofar as they employ their skills in starting new domestic enterprises or in modernizing domestic organizations. Mason defines technology transfer to include its transmission and assimilation by domestic organizations, without which there is a "technology transplant" but not transfer (Mason 1978a, 7). Turnover from one foreign enterprise to another does not contribute to transferring technology to the LDC; it only redistributes the benefits and costs of training between foreign firms. Although this is the main sequence, it can be bypassed. A foreign firm can transfer technology directly to a domestic organization—a supplier or a licensee—by training and technical assistance for its employees, without direct investment or turnover.

Turnover to the domestic economy not only increases the host country's ability to assimilate and use technologies new to it, but it compels foreign organizations to undertake more training to replace the employees they have lost. This provides a larger base of technical and managerial employees familiar with the new technologies and a greater potential for the diffusion of these technologies into the domestic economy via the turnover of employees.

> Recruitment and training of local personnel are considered to be one of the most important aspects of technological transfer by the MNC. This is more intensive as the subsidiary takes on more local staff at high levels of skill acquired largely in the MNC or with its help. There might be a multiplier effect through the mobility of labour: skilled workers leaving the subsidiary to work in national firms take scientific and technical knowledge with them to other sectors of the economy, thus helping to broaden the host country's capacity to absorb technology. . . . [T]he host country does not really benefit from the training acquired by its nationals on the management of MNCs, except insofar as there is a fairly high rate of rotation of staff of the subsidiary within the host country. (Germidis 1977, 17, 20)

The main vehicle of technology transfer is ultimately the turnover of trained and experienced managerial and professional employees from MNEs to domestic firms and other domestic organizations. It is conceivable for technology transfer to proceed without such turnover, with the continued expansion of MNEs as a share of industry and the economy. In fact this has happened, although more in developed than in less-developed nations (Canada, for instance, but also Singapore). But most LDCs preclude this route both through requirements or strong incentives for joint ventures with domestic firms, and through requirements for majority domestic ownership. In later stages of development, transfer without direct investment—by means of technical assistance to suppliers and via licensing becomes important.

Turnover indicates some absorptive capacity, but the rate of turnover cannot be closely related to absorptive capacity. High turnover indicates both some absorptive and shortage, an imbalance between supply of and demand for technology transfer-relevant skills. Low turnover, on the other hand, could mean almost any level of absorptive capacity, as long as demand does not exceed supply.

Turnover implies lateral entry (hiring of experienced workers for mid-level rather than bottom-rung positions). Net turnover from MNEs implies lateral entry into the domestic economy. This may indicate a rapid expansion in the industry or sector involved, with excess demand for trained and experienced workers, and with lateral entry needed to raise the attainable growth rate and therefore the attainable rate of new hiring and training of new hires. Lateral entries may contribute significantly to the training of new entry-level hires. Turnover and lateral entry to domestic organizations tend to be low in earlier stages of development, and to rise as an economy advances and its absorptive capacity grows (Kim 1980). Thus the potential rate of technology transfer and diffusion increases as an LDC advances.

Recruitment and training of managerial and professional employees, and their advancement to positions of responsibility, involve conflict between the interests of the LDC and of the foreign firm, to the extent that they weaken the parent firm's control over its affiliate, and/or to the extent they threaten control over the parent firm's technology. However, the latter can only occur through turnover that creates competition for the foreign affiliate.

Other things being equal, the more investment made, the greater the technology transfer potential; this is because a larger training base for turnover results. But the kind of investment matters. Investment in products and technologies for which local human resources are adequate involves no technology transfer—only capital movement. (Some additional technology or industry-specific training may also be involved. By increasing an already available supply, this training may lower the price of human resources for this type of investment). The more sophisticated the technology, the greater the amount of training required and hence the greater the potential for technology transfer via turnover. But complex technologies are likely to be capital-intensive, and create fewer jobs per dollar of investment. Thus there is a trade-off between the amount of technology-specific training per trainee and the numbers of workers trained. These then are the variables to be considered in estimating the maximum technology transfer:

1. Total amount of investment
2. Capital/labor ratio (trade-off between amount of technology-specific training per worker and number of workers trained)
3. Turnover rate (shortage of the skills trained; absorptive capacity of the domestic economy)
4. Trade-off between pay and turnover (payback period)

Of these variables, none is wholly determined by the MNE; numbers 2, 3, and 4 are much influenced by the absorptive capacity of the economy. Although the amount of a single new investment is determined by the MNE, it is the specific absorptive capacities that determine which investments are made in which products and technologies. (These investments add up to the total foreign investment.)

The only case in which absorptive capacity can be ignored is that of a large comparative advantage in mineral resources. This has been particularly true of oil and gas (into the mid-1980s). The difference between the cost of production of oil in Indonesia and its market price has been so large that it has been advantageous for foreign oil companies to build the required infrastructure from scratch and to employ a large number of expatriates, without which exploration, development, and production would not have been possible.

Determinants of Technology Transfer by Foreign Firms

Technology may be transferred in several ways: (1) through direct investment and production of new products, or in the use of new techniques; (2) by sale of rights to the product or to the productive technique (licensing of patents, trademarks, or franchising); and (3) in the sale of technology transfer services (technical services, management contracts, or turnkey plants). All of these involve some training, or human resource development.

Most firms prefer to produce and export from their home country, with the technology transfer limited to that technology embodied in their product and the training and technical assistance associated with sales of the product. Foreign investment and the accompanying technology transfer form a second choice. Firms may choose to invest abroad if they lose their comparative cost advantage in their home countries. They can supply foreign markets—and even the home market—more cheaply from overseas plants. Or they find that they can carry out some step of their production process more cheaply in overseas plants, and farm out their activities among plants in several countries. One of the main attractions of LDCs is cheap low-skilled labor. Firms may choose to invest abroad, whether or not overseas locations are cost-effective, if this is seen as the only way of retaining, or of entering, foreign markets. Firms are induced to move abroad by tariffs, quotas, outright import bans, or other measures that make it too expensive to serve overseas markets by exporting from their home countries. Some firms will invest in LDCs to obtain assured supplies of needed inputs. What some LDCs do not understand is that neither the foreign firm's home country, nor the LDC, can require the firm to transfer technology.

U.S. firms investing abroad have a strong preference for wholly owned subsidiaries, followed by majority ownership, and lastly, by minority ownership. Licensing of technology by U.S. (and also Japanese) firms is predominantly to affiliates, not to independent firms (Konz 1980, 61; Root, in Sagafi-nejad 1981, 122). Nonequity licensing typically is far down the list of preferences, although in countries where investments are subject to a great deal of uncertainty, a foreign firm may prefer licensing to equity commitment. The last choice would be technology transfer via technical services and management contracts, and the construction of turnkey projects. However, there are firms whose stock in trade is not technology embodied in a product or process, but exclusively the technical and managerial know-how of its employees. This is the case with engineering consulting firms, whose contribution to technology transfer involves no equity, but payment for which is (in some cases) partially an equity stake.

The willingness of a firm to transfer technology depends on its expected return from the transfer and on its opportunity costs. If the technology to be transferred is not proprietary in any sense (either legally, as with patents or trademarks, or operationally, as with know-how specific to the product or process being transferred), the opportunity cost of transfer may be considered to be zero, and only the expected return matters. Transfer is easier late in a product cycle than when a product still has a long economic life in its home country (this is Vernon's product cycle theory of location of production—Vernon 1966). The costs of transfer vary with the type of organization to which the transfer is made, being lowest to wholly owned subsidiaries, higher to joint ventures, higher still to independent licensees, and highest to government enterprises in centrally planned economies. Training costs vary by industry, being higher for machinery than for chemicals manufacture (Teece 1976, 24, 80).

Because the firm may be attempting to maximize expected returns on a global or at least on an international basis, its willingness to transfer a technology to a firm in a specific country may be conditioned on its retaining the ability to transfer profitably to other firms in other countries. Thus it will license a firm to use a process or manufacture a product only for sale within its own national market—and not for export—lest the licensee infringe on markets open to the licensor or to other potential licensees. Such restrictions limit the potential benefit from technology to the transferee.

If the technology is open—that is, not proprietary in any sense—there will be alternative transferors. Therefore, a transferor cannot deny a transferee access to export markets or restrict its use of the technology in other ways. Such open technology is not necessarily the latest but often is the most appropriate for an LDC. According to some observers, open technology sometimes supplies 60 to 80 percent of an LDC's technology needs, (Lentz 1980, 47).

To the extent that the technology is proprietary, the possessor is willing to transfer in direct relation to incremental earnings associated with transfer, and in inverse relation to the loss of proprietary rights or risk of loss of such rights and potential profits. A wholly owned subsidiary permits a possessor to retain the full returns on its technology and associated know-how; majority ownership dilutes that share but does not seriously increase the risks and uncertainties associated with technology transfer. Minority ownership, all the way to licensing of an independent firm, reduces a possessor's profit share and increases the risks of eroding its proprietary position. If the capital costs of implementing the transferred technology are high, the transferring firm may be prepared to trade off a reduced share of profits and greater uncertainty against reduced capital costs of transfer.

As Berle and Means indicated long ago, (1932) in *The Modern Corporation and Private Property*, control and ownership are often separate and weakly related matters. Control is the key element in profit maximization or satisfaction. Control can be maintained in several ways. Majority ownership is one way, with wholly owned subsidiaries being the extreme case. (The effect of minority ownership share on control depends on the concentration or dispersion of ownership of the local majority share.)

Majority ownership is not the only way of safeguarding profits or protecting technology. The same objective, at least so far as minimizing risk and uncertainty is concerned, can be accomplished via control of key managerial and technical positions. This is done by restricting the kind of training and technical assistance the workers receive, thus preventing them from acquiring the capability to use the technology themselves. Reluctance to advance nationals to senior managerial and technical positions may also be interpreted as reluctance to train, or subsidize, potential competitors.

Control may be maintained through transfer pricing, irrespective of ownership share, and often irrespective of the nationality of key managerial and technical personnel. Control via transfer pricing depends on the functional relationship of the affiliate (or joint venture or licensee), and the parent organization. To the extent that the affiliate produces output that is input to the parent firm, and if the affiliate has no alternative markets, then the parent firm controls the price that the affiliate receives for its output. To the extent that the affiliate (or joint venture or licensee) must obtain essential inputs from the parent firm, and has no alternative source for these key inputs, then the parent firm can control the price of these inputs. It can thus determine the allocation of profits, whatever the share of ownership (Svejnar & Smith 1984). In export industries, a MNE may maintain control through control of international marketing.

Control of key managerial positions widens the scope for controlling transfer pricing, but may not be necessary. Such control may be provided through licensing agreements even without any ownership share or management control. Thus one has to look at the character of the economic interrelations of the joint venture and the foreign firm with which it is affiliated as the character determines bargaining power and the allocation of profits. Ownership share and use of expatriates tells us something, but not the full story.

National Policies Affecting Technology Transfer

Host-country policies concerning foreign investment and transfer of technology are more important than product characteristics such as maturity in determining who supplies what technology and through which type of contractual agreement. In the absence of foreign-investment and technology-transfer constraints, local producers usually seek equity participation of the supplier of process know-how even if alternative, participation-free sources of technology are available. The acquisition of technology in a packaged form may also be preferred. . . . Technology suppliers are fairly flexible in their response to specific local conditions and constraints. (Cortes 1982, 265)

Ownership Rights

National governments affect the amount and type of technology transfer by restricting the ownership rights of foreign firms. If wholly owned foreign affiliates are not permitted, with local equity participation or joint ventures required, there may be an adverse impact on the volume of direct investment. The decline becomes pronounced when the foreign firm's equity share is limited to less than 50 percent. There is also an impact on the type of investment: the more complex and dynamic the technology, the greater the effect (Fund for Multinational Management Education 1978, vol. 2, p. 14). Proprietary technology is more affected than nonproprietary products and techniques. "One of the most serious restrictions which LDCs impose upon multinational corporations attempting to set up subsidiaries in their countries is that foreign ownership not exceed fifty percent. While these restrictions may achieve other goals, the LDCs should be aware that they may impose severe limitations on the supply of technology" (Magee 1977, 37).

The significance of regulations that deter direct investment on technology transfer depends on the LDC's ability to absorb technology in "unbundled" form, with only temporary and limited participation of the supplier, and on the willingness of suppliers to transfer technology without an equity stake, that is, via licensing and technical assistance. The impact of ownership restrictions is much greater for countries in the first stage of development; these have to depend mainly on direct investment and "packaged" technology transfer.

Import Content

Policies requiring foreign firms to reduce the import content of their output have acceleration of technology transfer as one of their objectives. These policies represent a demand by the host country for the local manufacture of products whose production involves technologies that the investor does not consider appropriate for the country (in the light of available factor supplies and costs, and/or in the light of available markets). Such demands for high-technology investment may require, from the investor's standpoint, too much capital investment in a country where capital is scarce and expensive and where low-skilled labor may be cheap. In particular, the demands might necessitate too high a training cost to develop the required indigenous human resources. They represent an infringement on management autonomy. Foreign firms already in the country may be discouraged from expanding their commitments, and potential new investors are deterred from entering the country.

The reference is not to intermediate versus advanced technologies. It concerns products whose appropriate technology of production is capital- and human-capital-intensive versus products whose appropriate technology is economical in terms of both physical and human capital. In most cases, the choice of appropriate technology is practically the choice of both an industry and a product whose production technology is appropriate for an LDC, rather than the choice of production technology for a given product. (The subject of *appropriate technology* is discussed in an appendix to this chapter.)

From the standpoint of an LDC, introduction of inappropriate technologies in this sense means a misallocation of scarce resources, capital, and highly trained labor into activities with high capital-to-output ratios (both in terms of physical and human capital). Thus, the scarcity of these factors in the rest of the economy is aggravated. This misallocation reduces the direct effects of foreign investment on employment. There is no a priori reason to believe that any conceivable indirect effects would outweigh the disadvantageous direct effects. The income multiplier effect for the capital-intensive technology would likely be smaller, at least so far as payroll is concerned. The ability of advanced-technology investments to generate large secondary employment via local production of inputs is speculative. Much of any potential secondary employment via production of inputs might have the same drawbacks in an environment where both are scarce. Nor is there any a priori reason why an advanced-technology product has a greater capacity to induce secondary employment (supplier) effects than a lower-technology output. The inducement of additional training must be weighed against possible deterrent effects on investment should the skills generated not be the "right" skills in the "right" industries.

LDCs hardly ever consider the potential deterrent effects on foreign investment of an aggressively pursued local content policy. In particular, they overlook the effect on firms whose input linkages make them vulnerable to such a policy. To what extent does such pressure preclude the development of potential exports by raising the domestic cost of production? To what extent does such pressure discourage potential investors (or additional investment by firms already there) because of their fear of becoming "hostages" who are compelled to produce items they can more cheaply import, or whose import from the parent company is part of the production and marketing strategy of the parent company? Are pressures sufficient to generate enough uncertainty about future policies so as to deter foreign investment?

This is not to say that pressure on MNEs to reduce their import content is always unwise. Some MNEs invest in an LDC mainly for the purpose of increasing their exports to it. Investment in an assembly/packaging/formulation facility, with parts, components, and materials imported from the parent company, may give an MNE two advantages, both of which increase imports. The first is a lower tariff on the import of inputs for assembly or final manufacture in the LDC; the MNE would pay a higher tariff if it did not have a local plant. The other is that having a local plant is likely to give the MNE a larger share of the local market than it could obtain strictly through exports of the final products. The LDC government may eliminate one of these "unfair" advantages by eliminating tariff concessions. The MNE may be reluctant to reduce import content, although some imported inputs could be produced economically in the local economy or for the local market, because the MNE may have otherwise excess capacity in plants in other countries.

Replacement of Expatriate Workers

Requiring a too-rapid replacement of expatriates is a second policy affecting technology transfer for certain products and production processes. Expatriates are very expensive, and MNEs have both an economic incentive to replace them as soon as native workers with comparable skills are available, and an incentive to train such workers. But a forced pace raises training costs, and lowers product quality, possibly to the point where export markets are precluded. Production for domestic markets then becomes too unprofitable, or must be protected by higher import barriers, thus raising the price of the product to domestic consumers. New entries are discouraged by the unfavorable cost and profit expectations.

One of the main jobs of expatriates is to implement technology transfer, in large part by training citizens (Helfgott 1973, 242–3). A distinction should be made between two expatriate functions: training local workers, and management. Limits on the number of expatriates engaged in the former function, or their too rapid replacement, reduce training of indigenous managerial and professional workers and the rate of technology transfer; Limits on the number of expatriates in the latter function increase uncertainty about the parent firm's managerial control.

Government pressure to replace expatriates compels MNEs to increase expenditures on training and related activities for technical and managerial personnel; replacing expatriates prematurely may further raise costs by reducing productivity. It also increases the risk that the firm will lose control over its proprietary technology. It may deter additional investment by the MNE already in place, and investment by other MNEs for whom expected profits are reduced and uncertainty increased.

It does not follow that there should be no restriction on the number of expatriates; it does not follow that the optimal level and rate of replacement of expatriates will be attained by MNEs without any regulation or pressure from the LDC. Delayed replacement of expatriates, even though the number involved is very small, reduces technology transfer significantly because delay denies host-country citizens the opportunity for experience in the most senior managerial and technical positions. This experience is critical for the successful transfer of technology to local firms and other organizations. The local citizens with experience in high-level positions constitute a prime group of potential entrepreneurs for technology transfer via formation of new domestic firms as well as through modernization of firms already in existence. In occupations and industries whose supply of trained and experienced workers is adequate in quality and quantity, exclusion of expatriates may not affect technology transfer, much of which can then be accomplished without direct investment.

Efforts to accelerate technology transfer through accelerated replacement of expatriates *and* accelerated introduction of new products and processes (these aims conflict with one another) in combination may have a magnified negative effect on technology transfer by deterring new investment. They particularly discourage investment in export industries, where internationally competitive costs are imperative, and where domestic markets do not limit scale and output. Apart from deterring foreign investment—an effect that is essentially impossible to quantify—these policies tend to bias foreign investment toward traditional products and technologies, whose requirements for expatriates are minimal, and products whose import content is minimal (processing industries) or are unlikely candidates for import substitution.

Investment Incentives

Investment incentives are yet another policy affecting technology transfer. Most LDCs encourage foreign direct investment in various ways. The most common policies include tax holidays and other methods of reducing tax liability, and reduced or zero import tariffs on productive equipment and other inputs. Subsidies in the form of factory sites, provision of utilities, and training programs are also common. The impact of such policies on technology transfer depends upon their effect on total foreign investment, and on the composition of foreign investment by product and industry as well as by production process. The tariff exemptions for capital imports tend to bias investment toward capital-intensive products and processes, which may involve substantial technology transfer but also inappropriate technologies (or products). The effect of tax subsidies depends on the composition of the tax system and the nature of the subsidies. Reduced taxes on profits will attract different investments than reduced taxes on production costs.

There is serious doubt about the value of investment incentives for foreign business. To some extent it is a negative-sum game played by most LDCs, with MNEs bargaining for the best deal but not changing the volume of investment, or perhaps not even changing the national location of investment (Mason 1978a, 111). Incentives are short-term in most cases, and politically uncertain, so that firms are unlikely to be swayed by them if the economic considerations of costs and markets are not satisfactory. Even if investment incentives are effective in increasing the amount of MNE investment, and even if they improve their composition by product and industry, they come at a cost. The government might have used the foregone revenue to stimulate investment and technology transfer in other ways (for instance, increasing the absorptive capacity of the country). An assessment of the consequences of investment incentives for MNEs by Thailand and by Indonesia with regard to impacts on technology transfer would be a very difficult if not impossible task, which in any case lies beyond the scope of this book. Suffice it to say that investment incentives are used, and that they have an unknown impact on technology transfer.

Trade Policies

Another type of investment incentive, the main one for foreign firms in many LDCs, is protection for domestic production from import competition through tariffs, quotas, and outright import prohibitions. This type of incentive attracts firms to produce for a protected domestic market, and forces foreign firms formerly supplying a national market through exports to establish productive facilities or write off the market. Clearly the larger LDCs are the ones for which protected domestic markets are a strong inducement for foreign investment. Protection from import competition, in combination with tariff exemptions for imported productive inputs, are conducive to profitable business. Thus trade restrictions encourage technology transfer via foreign investment. To the extent that direct investment is discouraged by other measures, such as restrictions on foreign ownership rights or domestic uncertainties, then the foreign firm exporting to an LDC still has the choice of retaining some earnings from sales in the domestic market by licensing a domestic firm to produce in its stead.

Two other policies have effects similar to those of a protective tariff. One is government procurement policies, which may reveal a strong preference for domestic firms. This is a powerful inducement for investment in industries with demand dominated by government procurement.

A second policy is counterpart trade—the requirement that in order to import inputs, firms must generate additional exports equal to the value of inputs they wish to import. This policy has effects that partially counteract tariff concessions, because it makes imported inputs more expensive indirectly. But on the other hand, like a protective tariff, this policy encourages replacement of imports by domestic production.

Industrial Property Rights

Numerous other policies may affect foreign investment and technology transfer even though that is not their primary purpose. One concern is the protection of intangible property rights (patents, trademarks, copyrights, and the like). Lack of adequate laws or of their enforcement deters both licensing to local firms and local production by an MNE, which would otherwise generate local know-how that could be transferred to competing producers. This lack tends to restrict technology transfer (whether by licensing, sale, or direct investment) to old technology. An effective patent system and a well-conceived and enforced system of standards are among the primary concerns of corporations engaging in research and development (U.S. National Academy of Sciences 1973, 24ff.).

Another concern is the protection of the MNE's more tangible property rights, including the risks of nationalization and procedures for compensation, and constraints on repatriation of profits and on royalty payments.

Where profits through share of ownership are not a factor (if the host nation does not allow foreign investment in particular industries, for example, transport or communications); or if the foreign company does not wish to invest in the host nation, for whatever reason (political uncertainties, or inability to produce in the host nation at internationally competitive costs), then the foreign company may sell the right to use a process or to produce a product or service. Its willingness to do so depends on two factors: the price it can obtain for its proprietary rights, and the safeguards provided in the host country to those proprietary rights. Patented technology requires adequate laws and enforcement of the patent rights of foreign firms. Know-how and goodwill associated with trademarks naturally require adequate trademark protection. Given adequate protection, licensing depends on the income thereof. The low ceilings on fees payable set by some LDCs, and/or the short periods over which fees are payable, do not make it worthwhile to transfer proprietary technology as long as there is any risk associated with it (piracy, for example), and/or as long as there is any opportunity cost (lost exports, for instance).

If the host country sets a ceiling on licensing fees, then the MNE reduces the services it provides accordingly, or obtains additional income by other means. The MNE may reduce technology transfer-related assistance; it may require the licensee to purchase imports from the MNE; it may insist on distributing the exports itself, subject to the MNE's transfer pricing controls; or the MNE may stay out of that national market. Most technology payments are made by affiliates and are subject to manipulation (Konz 1980, 80). "[H]ost governments can do very little *directly* to lower the price of technology imports. . . . The effectiveness of regulatory measures to lower technology transfer prices . . . is highly questionable. Even if they were effective, the opportunity cost in lower *availability* of desired technology can be high" (Root, in Sagafi-nejad 1981, 130). The United Nations' Joint Unit on Transnational Corporations concludes: "Intervention in the imports of technology which aim chiefly at financial and legal regulation without adequate technical infrastructure in the public and private sectors would defeat technology transfer objectives in the long run" (United Nations 1984, 36).

"The hard fact of the business world is that the most desired technology is rarely negotiable. Such technology is transferable only as an inseparable ingredient of foreign investments . . ." (Koldepp, in Sagafi-nejad 1981, 292). Baranson (1981, 5) notes a decline in interest among U.S. corporations in equity investment in LDCs because the risks are perceived as too high relative to expected returns. Thus licensing may be preferred as an equity-free way of obtaining additional returns from investment in technology in countries where the uncertainties of investment are large, provided the licensor's property rights are protected. Independent licensees (or joint ventures with minority ownership) are ways of minimizing risks. Nevertheless the range of technologies transferable at arm's length is more restricted than that available through direct investment.

Excluded Sectors and Occupations

The reservation of certain industries for domestic firms (and some for public firms only) constrains technology transfer by precluding foreign investment. Transportation, utilities, communications, and often banking and mining are restricted to domestic firms. Technology transfer therefore is limited to turnkey plants and, to a limited extent, to licensing. The amount of technology transfer is thereby limited, and new technologies precluded.

Exclusion of expatriate workers from some occupations may also restrict technology transfer in the products and processes for which these occupations are important.

The magnitude of these effects may be minimal, or may be large, depending on the adequacy of the absorptive capacity of the LDC with regard to the exluded industries and excluded occupations.

Competitive Conditions

A final consideration for foreign firms supplying LDC domestic markets is public policies affecting the right to produce under economically competitive conditions. Many countries have public firms producing in the same industry and in competition with private firms; the latter may feel uncertain about the terms under which they will have to compete. Public firms may be allowed or even required to sell at a price below the cost of production; they may have privileged access to subsidized credit or other subsidies that allow them to compete unfairly with domestic or foreign private firms. Public firms may not be required to make a profit in order to survive. Foreign investors will consider risks high in industries where they must compete with public firms, or possibly if they must depend on public firms as markets or suppliers. Just how high these risks are depends on the practices of government enterprises and on confidence in the consistency of government policies with regard to the performance of public enterprises. Public policies limiting investment in industries supplying the domestic market (that is, limiting the number of firms given investment incentives) paradoxically could encourage investment by reducing the uncertainties of competition.

Summary

Which kinds of technology (investment) are most affected by each of these deterrents? Restrictions on ownership and control are most important for proprietary technology, and particularly for advanced technology; they are of much less importance for open technology; here the firm's competitive advantage derives from superior management, marketing, and the like, rather than from unique technology. Firms with rapidly changing technology or products place more stress on majority equity (Fund for Multinational Management Education

1978, vol. 2, p. 14), in part because technology transfer is a continuing activity with high costs, but also because of the necessity of recovering the costs of innovation. Firms with diverse products are more tolerant of ownership dilution (Franko 1971, 73) because they are more likely to rely on a local partner for local marketing than firms with a limited product line. This is because the former firms compete more on a product differentiation basis, with a greater spreading of risks, and less on a cost basis.

Restrictions on control (whether through restrictions on ownership or on expatriates, or through domestic content legislation, or other regulations) further differentially affect export industry. An export industry must remain internationally competitive in order to survive and earn a profit. Firms supplying only domestic markets can be assured a return through protection from imports, and also through restrictions on domestic competition. However, protection from imports and from domestic competition derives from governments, whose policies may change, and hence do not offer the same assurance as derived from firm autonomy in product and process decisions, in sources of supply, and in markets.

Inadequate protection for patents again differentially affects transfers of new versus old technology, and may preclude the former altogether. Inadequate protection for trademarks affects products known for quality control and often for innovation (although by no means all trademarked products incorporate much technology), or with market advantages deriving from reputation if not from marketing organization. Restrictions on royalty payments also limit access to proprietary technology, hence to relatively new technology. Even when licensing agreements are not precluded, low royalties, uncertainties on transfer to home countries, or continuation of agreements reduce the flow of technical assistance and training that normally accompany a successful licensing agreement. In essence the patent or trademark owner will deliver less in return for a smaller payment (or higher uncertainty).

Countries in the first stage of development must depend primarily on direct investment for technology transfer, hence are most affected by restrictions on ownership and control. They lack the absorptive capacity to advance primarily through licensing or patent infringement. In more advanced stages of development, protection of patent and other industrial property rights can have a major effect on the amount and type of technology transferred.

Japanese and U.S. Businesses: Some Differences

A number of differences have been observed between the behavior of Japanese and U.S. firms investing in LDCs. Only those differences directly related to LDC investment incentives and deterrents and policies on technology transfer and human resource development are noted here. They will be kept in mind in conducting research in Indonesia and Thailand.

U.S. firms investing abroad have a strong preference for wholly owned subsidiaries, followed by majority ownership and minority ownership (Konz 1980, 61). Licensing of technology by U.S. (and also by Japanese) firms is predominantly to affiliates, not to independent firms. Nonequity licensing typically is far down the list of preferences. In terms of ownership, Japanese firms are much more accepting of joint ventures and minority positions than are U.S. firms (Mason 1978b, 16). Many explanations have been offered. As latecomers in overseas investment, Japanese firms have had to accept less favorable terms. Their reliance on Japanese expatriates allows them to retain control without majority ownership. To a greater extent than U.S. firms, they can rely on transfer pricing through tied sales or purchases as a vehicle for capturing profits. They are more adept at arranging for local partners who meet local requirements pro forma without affecting management.

Japanese investment in LDCs includes many smaller firms, whose entry and operation are facilitated by trading companies. The presence of smaller firms accounts in part for the greater Japanese acceptance of minority ownership in joint ventures. U.S. investment is dominated by large MNEs.

Japanese and U.S. investment in LDCs also differs considerably in industry composition (Yoshihara 1978, 198–201; United Nations 1984, 25–43, 63–74; Mason 1978b, 16–18, 52–58). Much Japanese investment is in traditional industries for which Japan is no longer a low-cost location, such as textiles and apparel. They have been called "relocated workshops" because they take advantage of cheaper, low-skilled labor in LDCs. Their decline in Japan provides surplus managers and technical workers available for overseas deployment— possibly a factor in the large number of expatriates. The technology of these industries is in no sense proprietary, so that there is less need to protect industrial property rights, hence less concern about maintaining control through majority ownership. (Mason reports that Japanese firms in high-technology industries are just as anxious as U.S. firms to have 100-percent-owned subsidiaries [Mason, in Sagafi-nejad 1981, 34]. Their profits depend on good management and low-cost production and market access. They produce largely for export. Their input requirements are compatible with supplies and costs in LDCs, and their old technology is readily absorbed by LDCs. By the same token these industries are more likely to be in competition with domestic firms. This at least was one pattern of Japanese investment in the past. Recent investment, however, resembles the U.S. and European patterns.

U.S. investment is much less in traditional industries; it is concentrated much more in modern industries and services, in products whose producers have some monopoly or oligopoly power—for example, "luxury" consumer goods. Less of it is attracted by cheap low-skilled labor, and more by protected local markets. A large part of U.S. investment is associated with patents and trademarks, and derives its profits in part from them.

Japanese firms typically bring in more expatriates than U.S. firms, and replace them more slowly. The slow replacement, rather than the high initial level, limits high-level training and experience for nationals. There are a number of partial explanations offered (Mason 1978b, 66–72). Some Japanese firms are in fact joint ventures of several Japanese companies, each of which wishes to have its own man on board. Retention of key managerial, technical, and professional positions by Japanese MNEs means they maintain control in the absence of majority ownership or in joint ventures with local firms. It is difficult to integrate non-Japanese into the Japanese management system. Other reasons given are language barriers, and a need for rapport that is satisfied by compatriots—rapport not only within the firm, but with embassy and business organizations.

U.S. expatriates are very expensive; their typical stays of two to three years are too short for optimum performance; and senior managers tire of constantly breaking in new personnel. Thus there are strong pressures to train local personnel to assume senior managerial and technical responsibilities. One might claim that U.S. firms use too few expatriates, and replace them too quickly. Because U.S. firms are unwilling to maintain a large number of expatriates abroad for long or indefinite periods of time, they often do not have this method available for maintaining control without turning to majority ownership. Therefore they are less likely to invest at all.

The Japanese practice of lifetime employment has been overstressed. In fact it is practiced in only one-third of the economy, and applies only to males. Turnover of U.S. workers after age 25 is little different from that of Japanese workers. However, there is a large difference in attitude toward turnover among larger, more modern firms. U.S. firms expect some turnover as a matter of course, and are prepared to hire workers who are leaving other employers, and even to recruit them. Japanese place much greater stress on continuity of employment and loyalty to the firm; a departing employee is looked at with suspicion both by the old firm and by the firms where employment may be sought. This difference in attitude is reflected in differences in recruitment of workers and in their training, advancement, and attitudes toward turnover. In one case, the worker is considered a member of the company; in the other, the worker is viewed as an efficient factor in production.

It is not our purpose to pass judgement on the merits of the differences in business practices of Japanese and U.S. firms. We only note that they exist, and that consequently they are differentially affected by local conditions and local government regulations. Japanese firms are more affected by restrictions on the number of expatriates and their length of stay, but less affected by restrictions on majority ownership, or by inadequate protection of industrial property rights such as patents and trademarks.

Investment attracted by a protected domestic market is influenced by the size or potential of the domestic market and by protection from import competition. Domestic competitive conditions matter. Investment incentives do not affect the choice of country and probably not the decision whether to invest or not. On the other hand, investment for export markets is not concerned with import competition but only with policies affecting cost of production—domestic content requirements, or import taxes on inputs. Investment incentives, if they lower production costs, might be a consideration.

Insofar as Japanese and U.S. investments differ in terms of local market versus export orientation, they are differentially affected by the several policies indicated in this section. Another way of looking at host-country policies is that they are not neutral between industries, or countries; we see that the set of existing policies determines, in part, which foreign industries will invest, and which countries will be represented.

Majority ownership is important to U.S. firms because it is equated with managerial control as well as with a majority share of profits. Managerial control is required in order to assure that profits are made. But the scope of managerial control, and hence the ability to control costs and assure profits, can be curtailed by means other than reducing the U.S. equity to less than 50 percent. The People's Republic of China (PRC), for instance, does not allow foreign firms the right to recruit labor, but provides workers on demand. Experience has shown that many of the workers proved to be unsuitable. It was difficult and time-consuming to replace them, and there was no assurance that their replacements would be any better. Thus, management in fact lost control over labor costs. This rules out major investment in export industry. Even if the PRC government were doing an excellent job of supplying the skilled workers needed by foreign firms, the introduction of an external decisionmaker adds to the uncertainty of business operations.

Most countries restrict the freedom of the employer to dismiss workers in some fashion. Requirements for minimum severance pay are predictable cost elements. On the other hand, forfeiture of the right to dismiss after a certain period of service either compels a firm to rotate most of its employees, with large additional costs, or confronts the firm with the prospect of a costly and unproductive labor force in the future. These restrictions are a deterrent to investment only insofar as they are greater in one country than in other countries where the firm might locate its subsidiaries. But they are greater deterrents to U.S. than to Japanese firms, given the differences in recruiting practices and attitudes toward turnover.

Appendix A: Appropriate Technology

The expression *appropriate technology* is often used to mean a technology that is low-skilled, labor-intensive, and capital-saving relative to the technologies employed in advanced countries with abundance of capital and highly skilled labor. Thus defined, the scope for appropriate technologies that can compete with the leading technologies employed in advanced countries is quite limited: agriculture, construction, some materials-handling, and short-distance transport nearly exhaust its possibilities. However, there is a much wider scope for appropriate technology. It may be differentiated in several ways from technology imported from a country such as the United States:

1. Adjusting production technology to local factor supplies and prices in order to minimize costs.
2. Adjusting to use of local materials. This differs from number 1 in that it involves substitution of a different input rather than a change in input proportions. For instance, shoe soles may be made of rubber instead of leather, or local fibers may be used in place of cotton.
3. Economizing on foreign exchange, that is, on imports of materials and equipment, even if this policy may increase costs of production.
4. Adjusting the scale of plant to local market size, for example, downsizing a plant.
5. Adjusting the product to local market conditions, for example, high transport and utility costs, difficulty with repairs and maintenance, lack of replacement parts, and low income.

None of these adjustments are foreign to the advanced industrial countries.

Appropriate technology is by no means always simple technology; it may be highly sophisticated. And appropriate technology developed in LDCs may well be quite sophisticated. An important example of appropriate technology developed in an LDC, which has been transferred to other LDCs, but in quite inadequate numbers, is the minisugar mill developed by M.K. Garg, an Indian engineer.

This type of appropriate technology, because it has not been developed or used by MNEs, is much less well known, and less accessible, than the standard technology of industrial nations. When technology is developed in an LDC, the problem of information and dissemination is aggravated. Given that the technology may not be simple, and that its diffusion involves establishing not a few plants in a large country but hundreds, possibly thousands, an LDC firm must confront the problems of regional supplies of managerial, professional, and technical labor, of essential infrastructure, and of supply sources. The conventional facilities built by multinationals or by engineering firms for host-country public firms circumvent this problem by concentration in a few large urban centers.

The concept of "appropriate technology" seems to assume that there might be a definite standard and method to determine the appropriateness of a given technology. But, in reality, there is no single standard as such. Thus, a foreign firm operating in a host country would regard the kind of technology that is economically feasible, including its practicality at the location, from the viewpoint of profitability. But the same technology may be viewed as less acceptable than other types by the host-country administrators for promoting the industrialization program they plan to implement, or less desirable from the viewpoint of existing socioeconomic conditions in that country at the moment. According to Santikarn, "the contribution of direct foreign investment should be judged by its ability to generate or foster human capital in the host countries. . . . An appropriate technology is one which . . . allows the creation of relevant skills both within an industry and in related industries for future industrialization" (Santikarn 1981, xii, 19).

The meaning of this is that MNEs in an advanced country such as the United States, whose technology may not be appropriate for some LDCs, given their resources and prices, cannot be expected to develop technology most appropriate to LDCs, and certainly not technology solely for the purpose of transferring it to selected LDCs (United Nations 1982, 22–23). The resources and institutions required to develop technologies appropriate to an Indonesia or a Thailand need not be developed in Indonesia or Thailand. The appropriate technologies may already have been developed in other countries with similar conditions; the resources and facilities to develop appropriate technologies may already exist, or may more readily be created, in other countries. There has been some transfer of "appropriate technology" developed in Third World countries, for example, in India, Thailand, and Korea, to other Third World countries. (Obviously this does not mean that Indonesia, or Thailand, should not have *any* such facilities, only that they need not have *all* needed facilities.)

Sometimes discussion of "appropriate technology" implies old or intermediate technology. But the NICs have advanced through hand-me-down industries but not through hand-me-down technologies. The textiles and later the steel, shipbuilding, and other industries that developed in Korea, Taiwan, and Brazil incorporated the most modern technologies.

Given the enormous gain in productivity at almost any factor–price ratio afforded by the latest technologies versus older technologies, talk of intermediate technologies is mere prattle in almost any industry that is expected to be competitive with imports, much less able to compete as an exporter in world markets. Or, as a U.N. official expressed it, "You don't select a technology; you select an industry. The industry selection is governed by the market" (U.S. National Academy of Sciences 1973, 39). The main exceptions are in agriculture, construction, and short-distance transportation and materials handling (International Labour Office 1973, 53; Spencer 1970, 149). Intermediate technologies are only feasible without large subsidies in nontraded goods and services. Even in this case, the technologies' main contribution is in employment generation, which may be small if the price elasticity of demand for their products is low. To the extent these technologies do generate substantial additional employment in the industry using the intermediate technology, they could be reducing employment elsewhere, if a higher price and inelastic demand mean a shift of demand from other, more modern industries (provided this shift is not away from imports). In some cases, an earlier generation of equipment, being phased out in the advanced industrial countries, makes economic sense for an LDC: The capital equipment is much cheaper, and such technologies typically are more labor-intensive than the latest generation, although not always low-skilled labor-intensive. LDCs are often ambivalent about the technology transferred by advanced nations as not appropriate to their needs, in particular their employment objectives. Countries happy to accept hand-me-down industries sometimes resent the introduction of hand-me-down equipment.

Appendix B:
Turnover

It is impossible to quantify the optimal turnover rate. No turnover implies no alternative demand for the skills trained, hence no technology transfer multiplier effect from the foreign to the domestic sector of the economy. A very high turnover implies an acute shortage, hence an excess domestic demand for the skills trained, and a large multiplier effect. However, it also implies high, possibly excessive labor costs for the MNE providing the training, because on the average, the trainee may not remain long enough with the training firm to pay back the cost of that training. The firm may have made a mistake in investment and the associated training expenditures, and may be moved to reconsider its commitment. One thing is clear, however: the optimal turnover rate from the viewpoint of the LDC is higher, perhaps much higher, than the optimal rate from the viewpoint of the foreign firm incurring the costs of training (and which only reaps a part of the benefits as long as the trainee remains with the firm).

It is not possible to quantify the maximum acceptable turnover rate because there may be a trade-off between turnover rate and rate of pay. An MNE may cut down turnover by raising pay and fringe benefits. Conversely, a lower rate of pay shortens the payback period, so that a firm can accept a higher turnover rate. There are even cases where turnover eventually increases the demand for a firm's products by providing knowledgeable employees to potential customers; profits from the additional sales should be offset against training costs.

It should be stressed that the turnover involving technology transfer is the net transfer from MNEs to domestic organizations, not that simply between MNEs, or from domestic organizations to MNEs. It should also be noted that zero turnover can mean that the domestic economy already has an ample supply of the skills required by MNEs, or a lack of demand—the domestic economy lacks the absorptive capacity, for whatever reason. In either case there is no technology transfer, but the first case calls for no corrective action, whereas the second calls for measures to increase absorptive capacity. The lack of absorptive capacity, or inability of the domestic firms to employ profitably the technology-specific skills provided by MNEs, may be a result of a lack of capital, of enterprise, of management, or possibly of markets. The MNE may have saturated the domestic market for its products, and there may be no prospect of exporting profitably.

The turnover of professional and technical workers from MNEs is contingent on either turnover of managerial and entrepreneurial personnel or on their domestic generation.

Turnover, and therefore the transfer of technology, depends on the firm (industry) specificity of the managerial and technical skills provided by the MNEs. Managerial skills tend to be less industry-specific than technical skills. Some skills, such as computer programming, although technology-specific, can be widely used throughout the economy because the technology is not industry-specific. Financial, marketing, and accounting skills likewise tend to be general rather than industry-specific. Hence one would expect higher turnover, and more rapid technology transfer, in such skills than in technical skills highly specific to particular industries (petroleum exploration, for instance). Of course the higher turnover can occur only if the other industries exist and are hiring. In this sense growth and development in almost any sector of the economy, involving more hires, will increase turnover rates in the more general skills and the technology transfer multiplier effect associated with turnover. Generally, skills in the service industries are more transferable across industries than skills in the goods-producing industries (other than materials-handling skills). Thus interindustry substitutability of technology transfer-relevant skills, and the number of domestic firms in the industry (or industries) within (or among) which the skills are substitutable, and their rate of growth, all influence the turnover rate. Without growth there can be little transfer.

References

Baranson, Jack. 1981. *North-South Technology Transfer*. Mt. Airy, Maryland: Lomond Publications, Inc.

Berle, Adolph A., Jr., and Gardiner C. Means. 1932. *The Modern Corporation and Private Property*. New York: Macmillan.

Cortes, Mariluz. 1982. The Transfer of Petrochemical Technology to Less Developed Countries. In Devendra Sahal, ed. *The Transfer and Utilization of Technical Knowledge*. Lexington, Mass.: Lexington Books.

Dahlman, Carl, and Larry Westphal. 1983. The Transfer of Technology—Issues in the Acquisition of Technological Capability by Developing Countries. *Finance & Development* (December).

Fellner, William. 1970. Trends in the Activities Generating Technological Progress. *American Economic Review* LX (March).

Franko, Lawrence G. 1971. *Joint Venture Survival in Multinational Corporations*. New York: Praeger.

Fund for Multinational Management Education. 1978. *Public Policy and Technology Transfer*. 4 volumes. Sponsored by FMME, Council of the Americas, U.S. Council of the International Chamber of Commerce, and The George Washington University (March).

Germidis, Dimitrios, ed. 1977. *Transfer of Technology by Multinational Corporations*. Paris: OECD Development Centre. 1969.

Gruber, William H., and Donald G. Marquis. 1969. *Factors in the Transfer of Technology*. Cambridge, Mass: MIT Press.

Helfgott, Roy B. 1973. Multinational Corporations and Manpower Utilization in Developing Nations. *Journal of Developing Areas* (January).

International Labour Office. 1973. *Multinational Enterprises and Social Policy. Studies and Reports*, New Series, no. 79. Geneva: ILO.

Kim, Linsu, 1980. Stages of Development of Industrial Technology in a Developing Country: A Model. *Research Policy* 9.

Konz, Leo Edwin. 1980. *The International Transfer of Commercial Technology—The Role of the Multinational Corporation*. New York: Arno Press.

Lentz, Alice, ed. 1980. *ASEAN–U.S. Cooperation in Science and Technology for Development in the ASEAN Nations*, Seminar report, 9 October 1980, Singapore. Sponsored by Fund for Multinational Management Education, ASEAN–U.S. Business Council, and ASEAN Chambers of Commerce and Industry, in cooperation with U.S. Council of the International Chamber of Commerce.

Magee, Stephen P. 1977. Multinational Corporations and International Technology Trade, paper presented at *Colloquium on the Relationship between R&D and Returns from Technological Innovation*. Washington, D.C.: National Science Foundation, 21 May 1977.

Mason, R. Hal, ed. 1978a. *International Business in the Pacific Basin*. Lexington Mass.: Lexington Books.

Mason, R. Hal. 1978b. *Technology Transfers: A Comparison of American and Japanese Practices in Developing Countries*. Pacific Basin Economic Study Center Working Paper Series no. 7, Graduate School of Management, UCLA.

Niehoff, Arthur H., and J. Charnel Anderson. 1964. The Process of Cross-Cultural Innovation, *International Development Review* 6. (June).

Novack, David E., and Robert Lekachman, eds. 1964. *Development and Society.* New York: St. Martin's Press.

Sagafi-nejad, Tagi, Richard W. Moxon, and Howard V. Perlmutter, eds. 1981. *Controlling International Technology Transfer: Issues, Perspectives, and Policy Implications.* New York: Pergamon Press.

Santikarn, Mingsarn. 1981. *Technology Transfer: A Case Study.* Singapore: Singapore University Press.

Singh, Vidya N. 1983. *Technology Transfer and Economic Development: Models and Practices for Developing Countries.* Jersey City: Unz & Co. (Division of Scott Printing Corporation).

Spencer, Daniel L. 1970. *Technology Gap in Perspective.* New York: Spartan Books.

Spencer, Daniel L., and Alexander Woroniak. 1967. *The Transfer of Technology to Developing Countries.* New York: Praeger.

Svejnar, Jan, and Stephen C. Smith. 1984. The Economics of Joint Ventures in Less Developed Countries. *The Quarterly Journal of Economics* (February).

Teece, David, ed. 1976. *The Multinational Corporation and the Resource Cost of International Technology Transfer.* Cambridge, Mass.: Ballinger Publishing Co.

United Nations. 1982. *Transnational Corporations and Their Impact on Economic Development in Asia and the Pacific.* Economic and Social Commission for Asia and the Pacific, Joint CTC/ESCAP Unit on Transnational Corporations, Bangkok.

———. 1984. *Costs and Conditions of Technology Transfer Through Transnational Corporations.* ESCAP/UNCTC Joint Unit on Transnational Corporations, Economic and Social Commission for Asia and the Pacific, Bangkok.

U.S. National Academy of Sciences. 1973. *U.S. International Firms and R, D & E in Developing Countries.* Washington, D.C.

Vernon, Raymond. 1966. International Investment and International Trade in the Product Cycle. *The Quarterly Journal of Economics* (May).

Wallender, Harvey W. 1969. *Technology Transfer and Management in the Developing Countries—Company Case Studies and Policy Analyses in Brasil, Kenya, Korea, Peru and Tanzania.* Cambridge, Mass.: Ballinger Publishing Co.

Yoshihara, Kunio. 1978. *Japanese Investment in Southeast Asia.* Monographs of the Center for Southeast Asian Studies, Kyoto University. Honolulu: The University of Hawaii Press.

2
The Local Environment for Technology Transfer: Indonesia

Economy

Indonesia remains mainly an agricultural country, with rice the dominant crop. Yields increased considerably between 1973 and 1983, by 4.1 percent a year for paddy and 4.3 percent for coarse grains. The agricultural share of the population has fallen below 55 percent, but this decline reflects more the limits of ability of the farm sector to absorb more population than expansion of employment opportunities elsewhere. Most of the growth in off-farm employment has been in the informal sector, mainly services. Industry accounts for less than 10 percent of employment. Nearly all manufacturing is for the domestic market, with only 4 percent of output exported.

This still traditional rural society is overshadowed by the oil and gas sector which, especially since the OPEC price increases in 1973, has come to dominate not only exports but also government revenues and investment resources. Other natural resources—minerals, forests, and fisheries—are of growing importance, but the annual output of forest products and fisheries is still far below sustainable yield capacity.

During the decade of the 1960s, GDP grew at an annual rate of 3.9 percent; during the 1970s this rate doubled, to 7.8 percent. This was primarily a result of the increased investment resources made available by the large increases in oil prices in 1973 and 1978. During the third five-year plan period, 1979–84, growth was 5.7 percent, and it is expected to be 5 percent during the fourth plan period, 1984–89. Domestic savings persistently exceed domestic investment, so that the immediate financial constraint on investment, even in a period of lower oil export earnings, is not resources but the ability to mobilize them.

With population growth at slightly over 2 percent, the 1970s showed a 70 percent increase in income per capita. Meanwhile, substantial change occurred in the structure of the economy. Manufacturing, transport and communication, construction, and utilities increased their share of GNP. Within manufacturing, consumer goods, which had accounted for over 80 percent of output at the beginning of the decade, accounted for just under half at the end, as intermediate goods and capital goods greatly increased their share. Agriculture, growing at a rate of 3.8 percent, shrank rapidly as a proportion of GDP, although it was still the largest sector by far (World Bank 1983, 86–87; Roepstorff 1985, 37–40).

Sixty percent of the population is concentrated in Java, whereas resources are scattered in the many other islands that comprise the far-flung archipelago nation. As a result, requirements for transportation and communication are great, and these facilities are quite inadequate in many areas. Deficiencies in physical infrastructure also exist (for example, in the power supply and other utilities) and these in combination with transportation difficulties hamper the development of resources and industries in much of the nation. Resettlement programs, designed to move excess population to the less densely inhabited islands, impose additional requirements for infrastructure and are limited by the poor quality of most of the soils not being cultivated.

Human Resources

Education and Training

The labor force is young, and poorly educated and trained. Twenty-eight percent of workers never attended school, another 30 percent never finished primary school. Only 19 percent had any secondary school education at all. These deficiencies in years of schooling are being reduced rapidly; deficiencies in quality of educational preparation will take much longer (AID 1982; U.S. Embassy, Jakarta, 1984, appendix VI).

Indonesia has chosen to expand its educational system rapidly at all levels. This has resulted in a shortage of teachers in spite of a large increase in their number. Given the small proportion of the population with even a secondary school education, this expansion has been at the cost of improvements in quality, which remains a problem at every level. Because teaching salaries are very low, few teachers devote full time to their teaching duties; most of them have other jobs. Education is compulsory through the sixth grade, and extension is being discussed, but only one-third actually complete the six years. Only 7 percent of those completing secondary school pass exams for university admission.

At the university level, only 14 percent of the faculty in 1980 had any graduate training, and the bulk of those with graduate degrees were concentrated in a few institutions. Most of the better-qualified faculty hold other jobs and devote very limited time to courses or students. Because most students arriving at universities are poorly prepared, and the quality of their education at universities is low with few exceptions, they come out with high expectations but without the preparation to realize them. Attrition rates are very high and the average length of time to complete university studies is very long—seven years.

Although there is a large number of secondary vocational and technical schools (some private, some run by various ministries), enrollments are inadequate, training time is too short, the quality of training is generally low, and the link with potential employers, weak. The shortage of qualified and experienced teachers is particularly acute in this area. Considerable training in scarce skills is conducted by foreign firms for their own employees.

Labor Market

The labor market, except in the largest cities, functions on a local, informal, personal contact basis. Information is lacking for a smoothly functioning regional, much less national labor market. A major problem is the lack of standards for classifying occupations, testing skills, and for designing and evaluating training programs. Because most training programs are of poor quality, and many are little related to the needs of the economy, many graduates end up in unskilled and semiskilled jobs. Modern firms conduct much training that may be called remedial. The lack of standards, and the lack of information and mobility, account for the existence of shortages in some skills in which workers who have been trained for them cannot get jobs. The problem of unemployed college graduates in a country that is very short of well-educated people is primarily the consequence of very poor training and specialization in fields unrelated to the needs of the economy. Another disturbing factor in the labor market is the unrealistic expectations of graduates at every level in a society where diplomas have been the passport to secure government jobs. But the government, as it modernizes, must place greater stress on functional responsibilities and competent performance.

Cultural Values and Attitudes

Indonesia is too diverse for easy generalization. The comments that follow on the attitudes and values affecting the behavior of the labor force apply primarily to Java, not necessarily to peoples from other islands and diverse cultural traditions. Because the problem of providing employment and generating new firms is primarily the problem of overpopulated Java (and Bali), our focus on the characteristics of the Javanese labor force is appropriate.

The labor force is predominantly rural, much of it with little experience of modern work organization, and with attitudes toward work shaped in the nurturing ground of village society. Among the rural farm population, what has been called "subsistence mindedness" is common, in part because the small peasants lack the resources to change crops and methods, to take risks; because they lack the necessary information; and because of a traditional suspicion of the new. "Development mindedness" is found mainly among larger farmers, who have the resources and better information to reduce uncertainties (Penny & Gittinger 1970, 264–67).

More generally, most workers give priority to solidarity in their work relations over efficiency, if they are concerned about efficiency at all. A "problem-solving" style of work is still only found among a better educated, largely urban minority (Douglas 1970, 242).

Indonesian (Javanese) workers are described as unwilling to make decisions, an attitude also understood as an unwillingness to accept responsibility and to exercise initiative (attributed by some to the autocratic lifelong role of the father in Javanese village society). As workers they are best at repetitive tasks that do not require flexibility, transmission of information, or reporting of results. They lack the planning ability to perform a series of connected tasks efficiently (Tsubouchi 1978). A muted suspicion of technology leads to a problem of self-confidence in technical areas. (A lack of interest in adopting more productive modern technology is particularly noticeable in Bali.) Workers do not maintain equipment; the notion of preventive maintenance is alien; it is not entirely a matter of know-how but of interest, and a lack of forward planning, with last-minute improvisation instead. Although there is a long tradition of craftsmanship, the concept of systematic standards and quality controls is not widely accepted, and performance is not recognized as a dominant determinant of pay and advancement. Workers are satisfied with a "more-or-less" approximation, and do not understand a "zero defects" mentality. In industrial activities there is a problem of poor safety awareness, perhaps in part because most Indonesians did not grow up in a machine culture, although this can be attributed in part to machismo.

Workers do not identify with the firm but with their social group; their society is structured by patron-client relations, although migration and urbanization are weakening these ties. They lack a strong work ethic, in the sense of valuing performance and understanding the relationship between their effort and outcomes for the organization. Workers are reluctant to accept supervisory discipline, and supervisors are often unwilling to exercise their responsibilities when workers are not performing their tasks acceptably (Willner 1969, 162–78; Tsubouchi 1978).

Indonesian managers also tend to have a short-term horizon and lack a problem-solving mentality. There is more concern with avoiding disagreements and with mutual acceptability than with performance. Managers tend to view technology transfer as a unilateral transaction rather than as a complex participatory undertaking involving major transformation of organizations and behavior.

Although the low productivity of Indonesian workers is related to the relative lack of work discipline and responsibility, ineffective management, and inappropriate incentive systems, as well as limited skills, a number of observers mention inadequate nutrition and poor health as contributory factors.

Most observers claim that there is a lack of entrepreneurial spirit among the Javanese, and that for this reason the Chinese have become dominant in business. However, it is not a lack of the spirit of enterprise but its direction. There is a basic cultural antipathy toward business and commercial enterprise, a stress on aristocratic values, a preference for values of learning and power, for diplomas and for government posts, a traditional reliance on government, and a demand for bureaucratic positions (Van der Kroef 1960). The Indonesian entrepreneur is alive and thriving in government. But aggressive behavior, individualism, and the dominance of materialist goals are not in keeping with Javanese cultural values. Among the private entrepreneurs, even the Chinese, there is a strong preference for short-term gains, and for speculation rather than the calculation and continuity required for business investment.

Indonesians of Chinese origin remain the largest underused entrepreneurial resource, particularly in the development of exports. Their commercial tradition, international links and experience, and better education have given them a share in Indonesian business out of all proportion to their numbers. But this success has also resulted in legal and extralegal discrimination against them in business, in joint ventures with foreign firms, in competition for government contracts, and in developing trade outside the main cities.

In the public sector, which plays a key role both in production and in the management and regulation of the economy, similar problems of responsibility and incentives are prevalent. The bureaucracy is characterized by top-down authority and communication, with decision making hampered by overcentralization and lack of feedback. Jobs tend to be graded on the basis of formal educational qualifications; public officials are hired on the same basis, rather than on functional responsibilities. Pay differences are not based mainly on performance, and promotion by seniority is emphasized over promotion by merit (Arndt & Sundrum 1975). Bureaucrats (and this applies also to entrepreneurs, managers, and workers) do not make a sharp distinction between their work and other aspects of their lives; nonservice goals tend to dominate service and organization goals in their priorities (Jackson 1978, 14).

Development Priorities

Indonesians, and foreigners concerned with Indonesia, tend to stress the enormous natural wealth of this vast archipelago. It is true that Indonesia has a wide variety and large amounts of mineral resources (oil and gas not least among them), and the extent of these resources has yet to be determined. And it has extensive forests, rich agricultural soils, and large fishery resources. But it also has 160 million people, and will have over 200 million before the end of the century. It is not clear that on a per capita basis Indonesia is that resource-rich. The fact that it is struggling to be self-sufficient in rice, its main staple food, is an indication that looking at resources without regard to the domestic need for them can lead to misleading conclusions. Stress must be placed on preserving the renewable resources, forest and fisheries in particular, so that less dependence is placed on depleting resources of oil, gas, and minerals as the engine of development. The depleting resources must be husbanded and their foreign exchange earnings used strategically if Indonesia is to finance the development of the human resources of such a large population. The development of renewable natural resources as an alternative basis for higher and rising incomes must be considered, once the low-cost depleting resources are skimmed off, and once domestic demand for them reduces the exportable output.

Indonesia's most abundant underutilized resource in the years ahead is an increasing supply of young, low-skilled workers with limited education. Nearly two million workers are expected to enter the labor force each year for the rest of the 1980s.

Official unemployment figures are not high but are meaningless in an economy in which most workers are self-employed. In fact a high proportion of the population is underemployed, and an even higher proportion seasonally unemployed. Most of the increase in employment in recent years has been in the informal sector, predominantly through self-employment, and services in particular, typically with very low earnings. Investments that will provide employment for them, and eventually an opportunity to advance in skill, receive high priority. However, the employment potential for such workers in further import substitution is quite limited. Remaining opportunities for import substitution tend to be in higher human-capital-intensive or high physical-capital-intensive activities.

Large-scale creation of additional jobs for low-skilled workers who will be entering the labor force in the years ahead also requires the development of additional export industries and/or large-scale expansion of the existing ones. "The basic constraint on Indonesian economic development is to be found not in the supply of investible resources but in the capacity to 'absorb' these resources and invest them productively" (Myint 1984, 45). The factors limiting investment in export industries must be addressed. Foremost is the relatively high cost of low-skilled, low-wage labor in Indonesia. Because of this labor pool's low productivity, and because of the high costs of doing business in Indonesia, the country is internationally uncompetitive. Opportunities for employment in low-skill labor-intensive activities certainly exist in the construction of domestic infrastructure, particularly transportation. A second limitation is the existence of even larger supplies of even cheaper low-skilled labor in India and China, as well as smaller but still large supplies in a number of other Asian countries. This international competition for labor-intensive low-skilled industry, and the limited growth of the market for some of these industries in the future, suggests that

Indonesia should follow a mixed strategy. Indonesia should encourage industries that use both of its cheap and abundant factors—low-skilled labor, and the natural resources in which it has a comparative advantage.

The problem of idle and unproductively employed labor is also one of regional imbalance. The densely populated islands of Java and Bali can no longer provide for a growing population on the land; migration to the outer islands can make only a small contribution to solving this problem. Solutions through job creation must be sought in the areas where the people live.

Indonesia's Current Development Plan

U.S. and Japanese contributions to technology transfer in Indonesia will be most valued and most welcome insofar as they complement and supplement the plans and priorities of the Indonesian government. It is with this in mind that we survey briefly the current (fourth) *Repelita* (1984–89).

A clarification is needed on the meaning of a five-year plan in Indonesia. *Plan* may be a misnomer. Indonesia is not a command economy. Even though a substantial part of nonagricultural activity (and some industrial crops) are in the hands of public firms, these public firms have a substantial amount of autonomy. A five-year plan indicates the government's sense of priorities for the years ahead, and because there are different interests, there is a large number of priorities. A plan's numerical components express hopes and expectations in some cases; in others, they express intended allocation of public resources. The plan is a reference point in shaping specific policies, in the allocation of financial and other resources, and in the use of incentives, licensing powers, and other regulatory instruments.

The 1984–89 plan depends heavily on the course of foreign exchange earnings from oil and gas, which is beyond Indonesia's control. It places substantial weight on foreign sources of capital and increased reliance on domestic private capital, which again lie outside Indonesia's control, although somewhat subject to its influence. Furthermore, there have been substantial policy changes—deregulation of the banking system and a sweeping tax reform among them—whose consequences are difficult to predict. In fact the subtitle of the plan document is "Policies and Prospects for Sustained Development under Challenging Conditions."

Furthermore, the plan is more in the nature of a political platform, offering something to most interest groups, than it is a coherent agenda that makes hard choices among contending priorities. As one observer remarked, there are so many bottlenecks in the Indonesia economy that it is difficult to set priorities. Yet in retrospect, with the sharp drop in foreign exchange earnings in the early 1980s, there were large cutbacks in investment projects, which in fact were a choice of priorities. The current plan scales back somewhat the ambitious objectives in the previous plan, but even so its assumptions on foreign exchange earnings from oil and on the availability of investment funds may prove optimistic. Its assumptions on the employment creation effects are surely much too optimistic.

The sequence of five-year plans reflects a consistent long-range view of the proper development sequence for Indonesia. The first Repelita (1969–73), which followed the Sukarno era, concentrated on rehabilitation of an economy in disarray. It targeted agriculture in particular, as well as critical infrastructure, and planned to spur development of supporting industries, especially fertilizers and related chemical industries and cement. Repelita II (1974–78) shifted emphasis to increasing employment opportunities, increasing productive capacity, especially consumer goods, and the reprocessing of raw materials into basic industrial materials and intermediate products. Additional impetus was given to input substitution in manufacturing. Repelita III (1979–83) continued the emphasis of Repelita II on increasing employment opportunities and the supply of consumer goods as well as a better distribution of consumer goods and social services. It shifted emphasis somewhat from import substitution to export promotion, and increased stress on processing of raw materials into intermediate industrial materials, and of materials into finished products. Both intermediate products and capital goods industries were to grow as a share of manufacturing output.

Planned budget expenditures of the central government have averaged about 22 percent of the GNP in the past decade, except for the 1981–83 period, when they were somewhat higher. The following sectors experienced a decline in their share of the planned budget: agriculture and irrigation, defense and security, and regional development (a slight drop). Sectors experiencing a marked increase in their share included education and health, and government apparatus. Industry and mining and other sectors showed no particular change (McCawley 1983, 10). Comparing Repelita IV with Repelita III (see table 2–1), the following sectors received a larger share of the development budget: mining and energy; education and youth affairs; health and family planning; housing; and business development. Sectors receiving a smaller share included agriculture and irrigation (a slight drop); transport and tourism; regional development; environment; and government apparatus (Rosendale 1984, 27).

Repelita IV (BAPPENAS 1984) gives priority to economic development with emphasis on agricultural self-sufficiency in food and on industries producing machinery for both light and heavy industry. It gives more attention to social development and development of other noneconomic fields "so as to make them mutually supportive for a sustained development." Development policies should be based on the three principles of equity, high economic growth, and a sound and dynamic national stability. Greater emphasis will be given to human resource development such as education, health, manpower, clean water supply, nutrition, housing, and human settlement.

Table 2–1
Sectoral Percentage Breakdown of Development Budget: Indonesia

Sector	Repelita III	Repelita IV
Agriculture and irrigation	14.0	12.9
Industry	5.4	5.4
Mining and energy	13.5	15.5
Transport and tourism	15.5	12.3
Trade and cooperatives	0.9	1.2
Manpower and transmigration	5.7	5.9
Regional development	9.8	6.9
Education and youth affairs	10.4	14.7
Health and family planning	3.8	4.4
Housing	2.4	3.8
Law	0.9	0.8
Defense and security	6.8	6.7
Information	0.7	0.6
Science and technology	2.0	2.3
Government apparatus	2.7	1.3
Business development	1.7	2.2
Environment	3.2	2.5
Religion	0.7	0.7
Total	100.0	100.0

Source: Cited in Rosendale, Phyllis. 1984. Survey of Recent Developments. *Bulletin of Indonesian Economic Studies* (April):27.

The growth rate of the GDP is planned to average 5 percent a year, versus the 5.7 percent achieved during the previous, third five-year plan (and the 6.5 percent planned for it). Agriculture and mining are to grow at a lower rate (3 percent and 2.4 percent, respectively), and manufacturing at a much higher rate (9.5 percent). The growth rates and changes in sectoral shares are given in table 2–2. The investment/GDP ratio is to rise to 26.3 percent, with private investment slightly less than public investment. Domestic private saving is expected to be slightly greater than government saving, rising at a 28 percent annual rate. Foreign savings should contribute 16 percent of the total, a reduced share, but an unchanged amount. Considerable reliance is placed on recent banking reforms—partially deregulating interest rates on loans and deposits, and limits on credits—as a means of attracting savings and financing investments.

Foreign private investment is to be directed particularly to capital-intensive and high-technology projects, as well as to those producing for export. It should provide additional employment opportunities, serve as a vehicle for transfer of

Table 2–2
Sectoral Growth Rates and Structural Changes: Indonesia
(*based on 1973 constant prices*)

Sector	Estimated Share in GDP, 1983–84	Average Annual Growth Rate, Repelita IV	Projected Share in GDP, 1988–89
1. Agriculture	29.2%	(3.0%)	26.4%
2. Mining	7.4%	(2.4%)	6.6%
3. Manufacturing	15.8%	(9.5%)	19.4%
4. Construction	6.3%	(5.0%)	6.3%
5. Transport and Communication	6.0%	(5.2%)	6.0%
6. Other Sectors	35.3%	(5.0%)	35.3%
Gross Domestic Product	100.0%	(5.0%)	100.0%

Source: BAPPENAS. 1984. *Repelita IV* (A Summary), p. 14.

know-how and technology in the shortest possible time, and should not cause environmental problems.

Government revenues from oil and LNG are estimated to grow by 16.7 percent a year, although revenues as a source of both foreign exchange and rupiah finance for investment are not expected to increase significantly, a more realistic assumption. A larger share of revenues will come from non-oil-and-gas sources. Much reliance is placed on the recent tax reform, with lower income tax rates, a 10 percent VAT, elimination of tax concessions for foreign investment, and measures to increase compliance, which should increase tax revenues. The allocation of the government development budget by sectors is given in table 2–1. Public investment will be directed toward labor-intensive and low import-content activities. Increased nonoil exports, however, are another objective of investment policy. Measures to encourage expansion of non-oil-and-gas exports include establishment of export processing zones, in addition to the banking reforms and the tax changes just mentioned.

The labor force is expected to grow by 9.3 million during the plan period, and the 5 percent planned growth rate (with its sectoral breakdown) is expected to provide this number of incremental jobs, with no increase therefore in the level of unemployment. This employment impact is considered by many as unrealistically optimistic. Expanded training activities include 650,000 to be trained by Department of Manpower Vocational Training Centers and development of apprentice training systems and in-plant industrial training. The government will train one million workers a year, and proposes that private institutions and industry train 3.8 million during the plan period.

Educational development will stress quality improvement in primary education, with a substantial increase in the teacher-student ratio. Junior and senior high schools will emphasize a large growth in enrollments, with enrollments and teachers in senior high school vocational programs increasing faster than the senior high school total (see table 2–3). Improvement in higher education will stress its role in the development of science and technology.

In agriculture, forestry, and fisheries, forestry is the sector with the largest growth target: an annual rate of growth of 7.1 percent for logs and 6.6 percent for forest products. Although the plan stresses diversification and expansion of secondary crops, the growth rate target for rice, 4 percent, is higher than for other food crops, plantation crops, fisheries, or animal husbandry. New lands will be brought under cultivation, existing water resources management systems will be rehabilitated and expanded, and new ones will be constructed.

In manufacturing, the plan emphasizes the development of small-scale industries to contribute to solving the employment problem. The plan envisions stronger linkages between small, medium, and large industries, including more subcontracting. The plan also stresses efforts to increase the exports of manufactured products. The annual growth target is 17 percent for metals and machinery industries, and 17.2 percent for chemical industries (including cement and paper). Slower growth targets are set for consumer goods industries. Products scheduled for rapid growth include industrial machinery, tools, agricultural tractors, railway rolling stock, ships and vessels, aircraft, fertilizers and pesticides for agriculture, cement, and pulp and paper. Manufacturing is expected to add 1.4 million workers, but 930,000 are to be absorbed by small-scale industry,

Table 2–3
Selected Targets: Primary and Secondary Education: Indonesia
(*thousands*)

Level	1983–84	1988–89
Primary school		
Students	28,869	29,380
Teachers	879	1,139
Participation rate (net)	97.2%	100.0%
Junior high school Students	4,713	7,738
Teachers	269	412
Participation rate (gross)	44.0%	65.0%
Senior high school Students	2,490	4,393
(of which vocational)	(552)	(1,113)
Teachers	162	280
(of which vocational)	(44)	(87)
Participation rate (gross)	25.3%	39.5%

Source: BAPPENAS. 1984. *Repelita IV* (A Summary), p. 59.

400,000 by miscellaneous industry (largely consumer goods) and only 70,000 by the remaining industries, including those just listed. Thus, the high-output growth target industries will use capital-intensive methods and contribute little directly to generating more jobs.

The plan calls for identification of industrial growth regions throughout the nation and the construction of industrial zones with facilities for industry. New miscellaneous industry will be concentrated as far as possible in industrial estates, which have the double objective of creating jobs where the need is great and of providing essential infrastructure at relatively low cost.

In the minerals sector, modest increases are projected for oil and gas, but a huge increase in coal output is projected, with the development of new mines. Substantial investments will be made in expanding electricity generating and transmission capacity, in land and sea transport, and in communications, all of which are currently bottlenecks for economic development.

Government Policies Affecting Foreign Investment and Technology Transfer

Ownership Rights

Since the 1974 amendment of the Capital Investment Law of 1967, only joint ventures with Indonesian partners have been allowed. (Exceptions are made for firms producing solely for export.) The Indonesian share of equity must be at least 20 percent initially, and must reach a minimum of 51 percent within 10 years. This policy, although widely circumvented by foreign firms of other nationalities, is believed to loom large as a deterrent to U.S. investments. The same law also tightened the conditions for employment of expatriates. Ceilings have been established by industry.

Expatriate Workers

Expatriates are typically to be replaced within three years in services, and within five years in industry. Suggestions have been made that the total number of expatriates should be reduced 15 percent per year. The "tax" for employment of expatriates beyond the number or time allowed was raised from $100 to $400 a month in 1983, but abolished in 1985 to eliminate misuse of revenues (intended for training replacements) and to encourage foreign investment. Expatriate employment requirements apparently were widely violated; enforcement has been tightened through a succession of "raids" on companies believed to have expatriates working without proper permits. Raids in the spring of 1984 netted nearly a thousand illegally employed expatriates, although in some cases the lack of permits was attributable to the dilatory behavior of the very ministry conducting the raids. Firms employing expatriates are required to establish training programs for Indonesians to replace the expatriates. Opinion is divided on whether restrictions are excessive—whether they raise costs and reduce technology transfer.

Perhaps as important as these official policies, which would greatly restrict foreign control if fully enforced, is a widespread attitude that foreign investment is grudgingly allowed for a time, while it is needed; that expatriates are tolerated while they are needed, but that both are temporary concessions to circumstances, to be dispensed with as soon as feasible. There is a widespread static view of the world: that every expatriate is depriving an Indonesian of managerial or professional employment opportunity, and that foreign business replaces domestic business. In addition to nationalistic hostility to foreign companies, there is a widespread ambivalence toward private enterprise (Palmer 1978, 54, 152–3; McCawley 1979, 65).

Domestic Content

Domestic content requirements have been pressed most vigorously in the motor vehicle industry, with 85 percent domestic content required for passenger cars and 100 percent targeted for late in the 1980s. Imports of complete motor vehicles are banned, as are imports of complete radios and TVs, a variety of textiles, iron products, batteries, and light bulbs. Effective protectionism as high as 718 percent on motor vehicles has been reported (*Economist* 30 April 1983, 100). The evidence on the consequences of these policies for technology transfer is mixed. On the positive side it has led mostly to in-house manufacturing, with independent suppliers being mainly other foreign firms, thus limiting technology transfer impact. On the negative side, foreigners are reported to be reluctant to invest in shipbuilding because they lack confidence in the government's policy on the use of domestic components (*Indonesia Times* 10 August 1983, 1). There are pressures also to use local products and local consultants in development projects.

Countertrade

A countertrade policy, instituted in 1982 in response to adverse trends in the balance of payments, is nevertheless likely to continue. It only affects those firms with contracts with the Indonesian government, or state enterprises obtaining foreign exchange financing from the state budget. It requires them, as a condition for the right to import materials or equipment, to purchase for export equivalent amounts of specified Indonesian commodities. This policy, which has engineering consulting firms trying to export tobacco, for example, is inducing U.S. banks to assume the role of trading companies for a fee. It reduces the value of the right to import materials and machinery and of the tariff exemption on such imports.

Investment Incentives

Tax incentives for foreign investors were eliminated in 1984. But the accompanying tax reform involving lower rates and simplification of the tax system should keep the tax burden of foreign investors from rising much (if at all) in the long run. It is questionable that tax incentives contribute much to the volume of investment, particularly when, as in Indonesia, incentives are predominantly (natural resources investments aside) for the domestic market. The tax reform is also intended to simplify enforcement and increase compliance, thereby reducing the scope for administrative discretion and taxpayer uncertainty.

Investment Licensing

Foreign investments must be approved by the Investment Coordinating Board (BKPM). BKPM screens applications for investment licenses in terms of their compatibility with national plans and priorities. It monitors technology, so that Indonesia does not become a dumping ground for obsolete and polluting technology. There is a ban on import of secondhand equipment. BKPM also limits licensing of foreign investments to avoid development of excess capacity, and to prevent entry of competitors in products for which it has already licensed foreign firms. These last procedures are described as complicated and ineffective (McCawley 1979, 38). They have not prevented excess capacity in many industries. Licenses are limited to 30 years.

There have been repeated reforms intended to simplify and accelerate the process of investment approvals. Obviously they did not work, or new calls would be unnecessary. In 1973 the government simplified the time-consuming procedures for foreign investment approvals, granting a reorganized BKPM more coordinating authority. In 1977 BKPM instituted a "one-stop service policy" under presidential decrees numbers 53 and 54. In 1983, President Suharto complained about the multiplicity of permits, reports, and levies that could slow down economic development. He demanded action to simplify the process of investment approval.

In 1985, investment projects were reported to require only 14 separate approvals, compared to 24 in 1984 and 39 before 1979. Even after an investment is approved, many legal and procedural difficulties remain to be overcome in obtaining land, electricity, and water permits from various ministries, regional and local authorities; import and purchasing licenses, and permits for foreign personnel; approved lists of capital equipment and operating inputs eligible for import duty concessions; and permanent operating licenses.

Regulation

When businessmen complain about the high costs and uncertainties of government regulation in Indonesia, they are not referring primarily to BKPM. An institutionalized, legitimized system of side payments is pervasive in Indonesia and is not a source of uncertainty or unacceptable costs. In some cases its availability helps reduce costs of business. Uninstitutionalized side payments, on the contrary, deter investment, raise costs by causing uncertainty, undermine government plans and programs, and are all too common.

We are not speaking of payments to individuals to expedite actions that would have been taken in due course, and that are widely regarded as legitimate supplements to low government salaries. The rights to these payments are clearly assigned, the schedule of payments is known, and the outcomes are assured. We are talking about payments to alter decisions. Such activities distort and pervert the interest of the government agency involved, as do payments in response to threats of adverse action (Anderson 1981, 113; Palmer 1978, 164–66). Many firms with long experience in Indonesia have learned to cope with the intricacies of regulation and to reduce their own costs and uncertainties. However, new investors view the situation as a strong deterrent.

Top government officials speak out against these practices. Reforms have been announced more than once to reduce the delays and uncertainties that raise the costs of investment and business operations, to reduce the number of public gatekeepers and the opportunities for bureaucratic gatekeepers' private gain at the expense of business and government's own interests. But the writ of government does not run far. The recent draconian measure of placing a Swiss company, Societé Générale de Surveillance, in charge of checking most of Indonesia's imports and exports, and putting half of Indonesia's customs officials on indefinite paid leave suggest that the government is serious about rooting out incompetence and corruption at least in the foreign trade area (*Economist*, 25 May 1985, 70–71). The problem is also reported to be acute in tenders, licensing, and taxation (the 1984 tax reform and plans to computerize the system are intended in part to reduce the problem in this area), and to be pervasive elsewhere.

Excluded Sectors and Occupations

Foreign investment is restricted in a number of sectors. In oil and gas it is limited to production-sharing contracts; in mining, to production-sharing or work contracts. Since 1975 foreign investment has been excluded from new forestry developments, although joint ventures can participate as logging contractors. Other sectors closed to foreign investors include shipping and ports, telecommunications, mass media, aviation, public railroads, atomic energy, and defense industries. Foreign involvement in banking, insurance, construction, legal, and accounting services is heavily restricted. Import and export licenses and domestic distribution are largely restricted to Indonesians. Land cannot be owned by foreign joint ventures—only by Indonesian nationals, and this discourages some agricultural processing investments. In a number of the areas above there could be substantial additional foreign investment and contribution to technology transfer, banking being an example.

Public firms are found in many sectors of the economy, even in agriculture. Although there is little complaint about their competitive practices by private firms in the same industries, there is some fear of the possibility of unfair competition based on privileged access to credit, markets, and tax privileges. The large role of public firms has been mentioned as a deterrent to private investment in cement and sugar, industries the government is encouraging to expand.

Industrial Property Rights

There is no patent law; one has been in process of enactment for a number of years. There is a procedure for registering patents with the Department of Justice. Nevertheless, it is the firm conviction of the foreign investor that there is no protection for patents. There is trademark legislation, but the uncertainties, delays, and costs of prosecuting violators are such that protection is of limited effectiveness. If a complainant wins a criminal case, civil action may be instituted for damages; however, the government journal was reported in 1984 to be five years behind. Counterfeiting and trademark violations are common; patent infringement is relatively rare.

Attitudes are a further source of concern. Indonesians prefer consumer goods bearing foreign trademarks. Understandably, this attitude is deplored by some Indonesian businessmen. It has been suggested by some government officials that foreign trademarks and brand names be banned. Clearly this attitude brings discomfort to firms whose trademarked products are being manufactured in Indonesia, whether by an affiliate or by an independent firm under license.

Licensing agreements are typically limited to five years' duration, with extension a possibility; allowable royalty payments vary by industry, and are typically 2 percent of net sales, which is the limit for tax deductibility. All agreements must be approved by BKPM. Given the limited size of the domestic market for many products at this time, the compensation allowed is often insufficient. However, firms have found ways of supplementing the payments; some government firms substantially exceed them; and most licensees are affiliates of the licensor, with alternative means of compensating the foreign firm. The limits on licensing agreements may however be a factor in the low proportion of licensees that are independent domestic firms. Pressure is applied to complete the process of technology transfer to licensees, curtailing the duration of agreements and payments. The only technology likely to be transferred under these circumstances is that near the end of its economic life in the industrial countries and which the licensee is not expected to be able to use to compete with the licensor in other markets.

There is inadequate appreciation of the fact that technology transfer is a very time-consuming, labor-intensive process. It is a costly undertaking for the transferor simply in terms of out-of-pocket costs, and often even costlier in terms of foregone opportunities for production and sale on its own account. Sometimes the transferor creates competition that will cut into its market share.

The insistence on buying technology cut-rate, when combined with restrictions on the use of expatriates and requirements for joint ventures with eventual minority ownership by the foreign firm (thus weakening or destroying control by foreign partners) assures that very little technology will be transferred which is not old or widely available.

References

AID. 1982. Jakarta Division of Education and Human Resources. *Detailed Analysis of the Indonesian Education Sector* (October).

Anderson, Benedict R. O'G. 1981. Looking Back. *Wilson Quarterly* (Spring) : 112–25.

Arndt, H.W. and R.M. Sundrum. 1975. Wage Problems and Policies in Indonesia. *International Labour Review* (November). 360–87.

BAPPENAS (National Development Planning Agency, Republic of Indonesia). 1984. *Repelita IV*, The Fourth Five Year Development Plan of Indonesia, 1984/85–1988/89 (A Summary) (May). Jakarta.

Douglas, Stephen A. 1970. Science and Technology and the Political Culture. In Howard W. Beers, ed. *Indonesia: Resources and Their Technological Development*. Lexington, Ky.: The University of Kentucky Press.

Economist (London). 1983. How Not to Develop. 30 April, p. 100.

Economist (London). 1985. A Custom of Corruption. 25 May, pp. 70–71.

Indonesia Times. 1983. Shipbuilding not Attractive to Foreign Investors. 10 August, p. 1.

Jackson, Karl D. 1978. Bureaucratic Policy: A Theoretical Framework for the Analysis of Power and Communications in Indonesia. In Karl D. Jackson and Lucian W. Pye, ed. *Political Power and Communications in Indonesia*. Berkeley: University of California Press.

McCawley, Peter. 1979. *Industrialization in Indonesia: Developments and Prospects*. Occasional Paper no. 13. Canberra: Development Studies Centre, The Australian National University.

———.1983. Survey of Recent Development. *Bulletin of Indonesian Economic Studies* (April) 1–31.

Myint, Hla, 1984. Inward and Outward-Looking Countries Revisited: The Case of Indonesia. *Bulletin of Indonesian Economic Studies* (August): 39–52.

Palmer, Ingrid. 1978. *The Indonesian Economy since 1965*. Padstow, Cornwall: Frank Cass.

Penny, David H., and J. Price Gittinger. 1970. Economics and Indonesian Agricultural Development. In Howard W. Beers, ed., *Indonesia: Resources and Their Technological Development*, Lexington, Ky.: The University of Kentucky Press.

Roepstorff, Torben M. 1985. Industrial Development in Indonesia: Performance and Prospects. *Bulletin of Indonesian Economic Studies* (April): 32–61.

Rosendale, Phyllis. 1984. Survey of Recent Developments. *Bulletin of Indonesian Economic Studies* (April): 1–29.

Tsubouchi, Yoshihiro. 1978. Indonesians at Work through Japanese Eyes. Discussion paper no. 99. The Centre for Southeast Asian Studies, Kyoto University (March).

Van Der Kroef, Justus M. 1960. The Indonesian Entrepreneur: Images, Potentialities and Problems. *The American Journal of Economics and Sociology* (April): 413–25.

U.S. Embassy, Jakarta. 1984. *Labor Trends in Indonesia* (Annual Labor Report) (May).

Willner, Ann Ruth. 1969. Problems of Management and Authority in a Transitional Society: A Case Study of a Javanese Factory. In Robert O. Tilman, ed., *Man, State and Society in Contemporary Southeast Asia*. New York: Praeger.

World Bank. 1983. *World Tables*. Baltimore: The Johns Hopkins University Press.

3

The Local Environment for Technology Transfer: Thailand

Economy

Thailand is still basically an agricultural country, with 70 percent of the labor force engaged in farming. Many farm workers, however, seek off-farm jobs during the agricultural off-season, with large-scale seasonal migration, particularly to Bangkok. Agriculture's share of the GDP declined from 40 percent in 1960 to 26 percent in 1981. Although rice production increased at a 2.7 percent annual rate from 1973 to 1983, and other grains (largely corn) increased at a 6.6 percent rate, yields declined marginally in the former and increased marginally in the latter. Other crops have grown rapidly as farming has diversified, although nearly 60 percent of crop land is still in rice. Growth through expansion of agricultural land area is over. Commercial forest resources, teak in particular, have been depleted, as have been coastal zone ocean fisheries.

Manufacturing has increased its share of the GDP from 13 percent in 1960 to 21 percent in 1981 and currently employs some 10 percent of the labor force. It has diversified from its traditional concentration in agricultural processing activities into textiles and apparel, consumer goods, chemicals, and transportation equipment. It is in the process of creating a heavy industrial base centered on natural gas, initially including fertilizers and petrochemicals.

Exports (mainly agricultural products) are about one-quarter of the GDP, but again there has been diversification from the traditional rice and rubber into cassava, corn, sugar, seafood, and canned fruit. Tin is the main mineral export. Manufacturing is accounting for an increasing share of exports, now about one-third, mainly textiles and apparel, plus gemstones and integrated circuits. Petroleum is the principal import, but development of Thailand's natural gas supplies, which first came on stream in 1981, promises to reduce this import bill substantially.

The physical infrastructure was very substantially improved during the

1960s and 1970s. There is now good transportation throughout the kingdom, and an adequate supply of power and other utilities. Communications have also greatly improved, although the telephone system remains inadequate. New port facilities will be built southeast of Bangkok. Problems remain with traffic congestion and seasonal flooding in Bangkok. On the whole, the physical infrastructure is no constraint on substantial further growth. In some parts of the country there is need for new or expanded water management systems to increase the cultivated area under irrigation and to intensify Thai agricultural production.

Human Resources

Education and Training

Although there are shortages of particular skills, and quality shortcomings in some areas, there is adequate supply of labor capable of learning the skills necessary to foster new technologies or industries. Projections for 1986 show that 68 percent of new entrants in the labor force will have a primary education; 13.8 percent will have a secondary education; 12 percent, a vocational school education; and over 6 percent, a college education (Hongladarom & Charsombut 1984, 8). If one excludes agricultural employment, a sector having low educational attainment, the labor force is reasonably well educated with a good supply of skills. There are large differences in these respects, however, between the Bangkok area and "upcountry," especially the Northeast.

In 1980, 96 percent of primary school-age children and 29 percent of secondary school-age children were enrolled. Thailand in 1980 had approximately 247,000 university graduates, including 22,549 with masters degrees, and 2,698 with doctoral degrees. It had a total of 55,790 scientists and engineers and 68,500 technicians, as well as 161,500 craftsmen. Currently, Thailand is graduating each year between 1,000 and 1,500 each in natural sciences, agriculture, and engineering. The capacity for producing masters and doctoral graduates is expanding. Doctoral programs in the life sciences were established at Mahidol University in the 1960's, and doctoral programs in engineering and science are beginning at Chulalongkorn University and King Mongkut Institute of Technology, respectively. More than 100,000 Thais hold degrees from U.S. universities. Graduate education in public administration, business administration, and economics is well established at Thammasat and Chulalongkorn Universities. Together with the expansion of higher education, there has been much improvement in quality of faculty, facilities, and training. There are problems: the quality of some universities far from Bangkok is poor, and it is difficult to attract and retain good faculty. Some of the education tends to be too theoretical, being remote from Thailand's current needs. With the establishment of two open universities with huge enrollment, whose graduates are of uncertain quality, the country is producing a growing surplus of college graduates in most fields.

Cultural Values and Attitudes

Thais are reported to be good with machinery and heavy equipment, and to have no problems in mutual accommodation of modern technology and traditional culture. Thai males, however, have a strong preference for government employment, and the more highly educated they are, the larger the proportion working for government and public enterprises. The males, much more than the females, tend to have a fatalistic worldview derived from Theravada Buddhism; this weakens their perceived relation between effort and achievement. In a hierarchical society with status subject to change, they seek education, but for the diploma and the status it confers. Women are more likely to think instrumentally and to relate goals to performance. Quality control is a problem with males but not with females. There is however a receptivity to change, responsiveness to economic incentives, and a capacity for critical self-examination (Wyatt 1975, 125-50; Kirsch 1975, 172-96).

The individualism characterizing Thai social structure has some implications for transfer and diffusion of new technology. Individualism expresses itself in a proprietary attitude toward skills and knowledge, and in a reluctance to share as contrasted to a willingness to employ them. A study of adoption of new technology among farmers reports that farmers rarely informed fellow farmers of what they knew about new agricultural technology. More stress needs to be placed on one-to-one contact, and less reliance on the effectiveness of a diffusion process (Pontius 1983, 104). Farmers will individually seek out information and use it. And they readily learn and modify their practices from observation. A large food firm reports an "aura" effect among farmers working in proximity to its contract farmers who received technical assistance from the firm.

In an industrial context, skilled workers are reluctant to serve as tutors to inexperienced workers, nor do they feel obligation to the organization that provided them with their skills. Without specific provision for training responsibilities, informal learning on the job may be handicapped. And workers with skills have no compunctions about leaving for a better job. Management in turn is quite willing to hire at all levels, but is reluctant to dismiss workers.

Thais avoid involvement or commitment. They have little sense of obligation (Ayal 1969, 535-49). They are used to working together, but not as a group; they dislike group activity. Their loyalties are to individuals, and to some extent to their professions—but not to organizations. Working together is limited to one-to-one interaction between adjacent levels of authority and responsibility; the efficiency of multilayer and extensive horizontal coordination can only be achieved step by step. These attitudes create problems for management.

Horizontal links are limited; the organization of society is vertical. The

reciprocal patron-client relations that characterize Thai society are not primor-
dial relations, however, but are based on mutual interest, and are subject to
change. What has been described as "authority without autonomy" of the
patron needs to be considered in promoting technology transfer and diffusion,
particularly in agriculture, where alternative structures of influence may be
lacking (Isarangkun & Taira, 1977; Hanks 1975, 197-218).

The behavior of managers is influenced not only by the basic social structure
but by the prevalence of family firms. At the top management level, kinship and
personal relations govern decisions. At lower levels of management, hiring and
advancement are largely on the basis of preparation and performance. Seniority
is not a major consideration.

The supply of entrepreneurs is not a problem. Thai males have traditionally
preferred administrative and professional positions to entrepreneurial and commercial roles, relegating the latter to Thai women (who are disproportionally
represented in them), and to foreigners. For well over a century, Thailand has
welcomed foreign entrepreneurs as well as skilled workers. Foreigners and
Thai-Chinese are prominent in business and industry. The distinction between
government and business roles is not clear-cut because each sector is deeply
enmeshed in the other's activities and functions.

The organization and functioning of government reflects the nature of Thai
society. Government, as in Britain, is plural. Bureaucrats see their office as a
personal domain rather than a public responsibility. Hierarchies of authority are
loose, functional boundaries are fuzzy and permeable, and power is neither
monopolized nor permanent (Scott 1972, 574ff.). These characteristics lend
public administration some flexibility on the one hand, and unpredictability and
uncertainty on the other.

Major Problem Areas

Thailand faces two major interrelated problems: growing unemployment and
regional imbalance. The peripheral areas, the Northeast in particular, experience
low incomes and high rates of underemployment.

Total employment is about 21.5 million, of whom 5 million are seasonally
unemployed and 4 million underemployed. Open unemployment is only
3,000,000, a statistic that is meaningless in a society with a work force mostly
rural, agricultural, and self-employed. Open unemployment is concentrated in
the cities (Bangkok in particular), and among the more educated work force,
including 60,000 university graduates (Eoh 1984, 9; Hongladarom & Charsombut 1984, 17-34).

The unemployment problem will grow, as 700,000 workers are expected to
enter the labor force each year for the rest of the decade. Most of them have

limited education and live in rural areas. But the urban educated unemployed could pose a threat to social and political stability. The number of university graduates unemployed rose from 37,000 in 1980 to 60,000 in 1984. The problem is especially acute with graduates in education, the humanities, and social sciences. Most graduates prefer jobs with government, particularly state enterprises, but government has slowed down the rate of increase in its employment to 2 percent. The problem is not reluctance to accept jobs outside one's field of specialization; a high proportion of employed college graduates have done this. Nor is it just a problem of specialization in the wrong field. Demand is inadequate for the available supply.

The surplus of educated Thais represents resource waste as well as a social problem. The question is how the surplus can be turned into a resource, into an asset for economic growth. The surplus will have several beneficial effects on the labor market. The large disparity between government (and university) salaries and business salaries could be substantially reduced; that, plus the unavailability of alternative employment, should improve the quality of government hires, and facilitate retention of the abler faculty. The present scarcity, and consequently the high pay, of senior experienced managers will gradually disappear. Thus professional and managerial manpower is likely to become cheaper, both relative to average wages, and to such manpower in other nations. Industries with higher ratios of management and professional and technical labor should increase their competitiveness and gain in comparative advantage. In many cases they require export markets to achieve economies of scale with Thailand's current limited domestic market.

The extensive investments of the past two decades in economic and social infrastructure can accommodate new industries that make substantial demands on such infrastructures. The remaining difficulties involve regional distribution—improving infrastructure in some cases, but primarily in building up regional centers far from Bangkok to a critical mass that can attract new industry and absorb most of the rural-urban migration from their hinterlands.

The problem of unemployment cannot be solved in Bangkok. Jobs have to be created close to where much of the rural population is located. Whereas economic growth will absorb increasing numbers of educated workers (the demand for which is in large part a function of per capita income), growth will not absorb the increasing numbers of low-skilled workers in rural-agricultural areas. Regional centers must be improved to the point where they can attract new industry and absorb much of the rural-urban migration from their hinterlands, incidentally attracting some of the educated unemployed who crowd Bangkok. Because Thailand's farm population is some 70 percent of the total, it is difficult to conceive of policies that maximize income and employment opportunities for the rural-farm population that do not maximize national economic growth as well.

The Fifth Five-Year Plan

Thailand's first four five-year plans, starting in 1962, were a poorly interrelated combination of grand objectives in broad strokes and overdetailed programs for different ministries and sectors. Although Thailand grew at an average rate of over 7 percent over this period, the plans did not have much impact on the pattern of growth. The first stressed construction of essential infrastructure; subsequent plans stressed industrial growth through import substitution; and in the last half of the 1970s emphasis began to shift to export expansion. The fifth plan (1982-86) differs from the earlier plans in that it attempts to integrate the macro objectives and the micro plans and projects. It concentrates on major projects rather than attempting to specify plans in minute detail. And it also aims at improving the macro management system (Faith 1983, viii-xi).

In a system that relies heavily on private investment and enterprise, and whose government is far from monolithic, the fifth five-year plan should not be regarded as a definitive comprehensive blueprint for the economy but only as an indicator of a sense of priorities, of the corresponding direction of investment incentives, and of the intended allocation of public expenditures, including some specific major projects.

It is somewhat more conservative than its predecessors, aiming at a 6.6 percent annual growth rate of the GDP. It promises fiscal austerity, reduction in domestic consumption as a share of GDP, removal of subsidies to various state enterprises, higher taxes, and higher prices for some loss-making public utilities. It places more emphasis on rural development and on the elimination of poverty than previous plans, claiming that these objectives rather than economic growth alone are primary. Development of the poverty-stricken backward rural areas implies decentralization of government management to the village level, as well as close coordination at every level between several ministries. For development of the more progressive rural areas, the government is relying heavily on mobilization of private capital and on private enterprise, technology, and marketing ability.

The plan views the 1982-86 period as a crucial transition from an agricultural-based to an industry-based economy. A shift is seen in the composition of exports toward an equal share of manufactured and agricultural products, and the diversification of industry to include the development of basic industries.

In industry, priority is given to labor-intensive industries in which Thailand has a comparative advantage, that is, those with export potential. The policy thrust for investment incentives will shift away from import substitution to export expansion. At the same time, consideration will be given to increase local content. The Ministry of Commerce has established a master plan to expand exports by 12 to 15 percent a year for the three years 1984-87 in nine product groups: flour and flour products; fresh and canned fruit; frozen and canned meat; vegetable oil; rubber, plastic and leather products; textile products (especially apparel); gems; electrical and electronic goods; and construction materials and sanitary ware.

The largest project by far, which is the basis for changing the industrial structure of Thailand in the future, is the Eastern Seaboard development, centered on natural gas. It will involve petrochemicals, fertilizers, and an integrated steel mill. By its nature it is capital-intensive, but it will replace substantial imports, especially oil and petrochemicals. In addition to plants, there will be development of infrastructure: deep-sea ports, railroads, and power. Also to be developed is a nearby light industrial zone to take advantage of the infrastructure investment for heavy industrial development. This development, which will take more than one five-year plan period to complete, will also help reduce the overconcentration of industry in the Bangkok area, but will not shift it to the more distant regions with large underemployed populations.

As an integral part of the focus on rural development and reduction of poverty, more stress will be placed on location or development of industries in areas such as the Northeast. Although decentralization from Bangkok has been a consideration for some years, industry promoted by the Board of Investment (BOI) is more highly concentrated in the Bangkok and Central Thailand areas than other industry (56 and 85 percent of BOI-promoted factories, versus 21 and 41 percent for all factories, as of 1979).

The sectoral targets are somewhat more modest than those of the fourth plan. Planned annual rates of growth are given below:

Sector	Growth Rate
GDP	6.6%
Agriculture	4.5
Manufacturing	7.6
Mining	16.4
Imports	7.0
Exports	10.0

Gas production is to reach 525 million cubic feet per day by 1986, a target that does not appear attainable at this time. Oil imports are to decline by 3 percent a year. A shift from the previous plan is the increased reliance on taxation. Government revenues are to increase 22.3 percent, and expenditures 20.3 percent, averaging respectively 16.7 and 18.2 percent of GDP. Total borrowing is planned to reach $10.5 billion, of which $2.9 billion is to be from foreign governments and multilateral development banks. Specific forecasts are presented in table 3–1.

Government Policies Affecting Foreign Investment and Technology Transfer

Ownership Rights

Thailand allows 100 percent foreign-owned firms, particularly when over 20 percent of the output is for export, although joint ventures and a significant Thai equity share are strongly encouraged. Thai participation in equity is a consideration in decisions to provide investment incentives.

Expatriate Workers

Expatriates must have work permits. Industry-specific guidelines determine the number of expatriates allowed per firm. No U.S. firm interviewed complained about the rules or their administration, or indicated that it employed the maximum number allowed. Some Japanese firms, however, regard the number of expatriates allowed as too small. There appears to be no pressure to phase out

Table 3–1
Thai Economy: 1981 and 1986 Plans
(*current bahts*)

	1981	1986
GDP (*million bahts*)	817,000	1,859,000
Per Capita Income (*bahts*)	17,204	35,692
Labor Force (*thousands*)	23,756	27,505
Employed	23,495	26,955
Unemployed	261	550
Commodity Exports (*million bahts*)	162,627	444,952
Agricultural Products (*ratio*)	48%	43%
Manufactured Goods (*ratio*)	29%	42%
Commodity Imports (*million bahts*)	229,877	528,536
Commodity Trade Deficit (*million bahts*)	67,250	83,584
Percent of GDP	8.2%	4.5%
Total Balance of Payments Deficit (*million bahts*)	53,014	44,529
Ratio in GDP	6.5%	2.4%

Source: Royal Thai Government, Office of the Prime Minister, National Economic and Social Development Board. 1982. Executive Summary of Fifth Five Year Plan. Bangkok.

expatriates, and certainly no deadlines for their replacement by nationals. Expatriates working before 1972 can remain for life. Some U.S. expatriates have chosen to become nationals, but the number is unknown.

Domestic Content

Only the motor-vehicle industry has been required explicitly to replace imports with domestic products. (As indicated in the next section, a domestic content policy has been implemented implicitly by other means in the pharmaceutical industry.) The requirement that 50 percent of the value of motor vehicles be of domestic origin does not appear reversible, despite a wide awareness of its effect in greatly increasing the cost of cars. However, the BOI can terminate the eligibility of specific products for import duty reduction or exemption whenever it judges that the item can be produced domestically.

Investment Incentives

The BOI administers the Investment Promotion Act. The act provides a variety of subsidies and services for domestic and foreign investments that BOI considers important for the development of Thailand. Investments that do not meet its criteria can still be implemented, but receive no privileges. The BOI stresses agroindustries, labor-intensive industries, and export industries. Project eligibility characteristics include minimum investment amount, production for export, amount and source of capital, nationality of shareholders, (at least 40 percent local equity is preferred), use of domestic inputs, nationality of management, training and employment of manpower, and location outside the Bangkok area.

Privileges that may be granted include freedom from the following: nationalization; competition by new state enterprises; price controls and state monopolization of the sale of products of the type made by the promoted enterprise; and government import of competing products. A promoted enterprise may own land required for its operations, export its products, employ aliens within limits set by the BOI, and has wide latitude in remitting foreign currency. It may be exempted or granted reductions in import duties and business taxes on machinery required (and for one year on raw materials). It may be freed from corporate income tax for three to eight years, and from tax on dividends paid shareholders and royalties and fees for a variable number of years. Firms locating in an Investment Promotion Zone may receive additional tax reductions (American Chamber of Commerce in Thailand 1982, 52-53). As incentives for export, firms are refunded as tax credits their indirect tax payments, import duties, business and excise taxes on the goods exported. However, the fact that most U.S. investment is not promoted raises doubts about the effectiveness of, or need for, these incentives.

The BOI has instituted a one-stop investment center to deal with a major complaint of investors: bureaucratic red tape and slowness in decision making.

Export taxes, which formerly were a significant deterrent to investment for export, have been eliminated for most goods and now account for only 2 percent of tax revenue.

Import duties are a major source of revenue for the Royal Thai government (RTG) but they are also a major incentive for foreign investment in industries producing for the local market. In many cases the domestic market is too small to offer least cost scale economies, hence domestic production is only viable with protection from imports. Most Japanese and U.S. investments are for the local market. Changes in import taxes on specific items have been used to offer protection for specific firms, and some foreign firms in Thailand complain that domestic competitors have been given an unfair advantage without forewarning. Import duties were raised in 1984 on a large number of drugs in order to protect domestic producers and to induce domestic production. This is by no means the only industry investment that is encouraged by protection from imports, but it appears to have been targeted recently.

Industrial Property Rights

A patent law took effect in 1979. It is too early to judge the effectiveness with which it will be enforced. Trademarks have been protected by law for many years. However, there is considerable piracy of trademarks—both of locally produced goods (especially drugs), and of imports. Producers of spurious or substandard drugs are subject to heavy fines and imprisonment. Enforcement of trademark legislation (presumably also of patent legislation, but it is too soon to judge this) is possible, but protracted, uncertain, and costly. The penalties are so light and delayed that there is no effective protection. This is a problem for U.S. investors who serve a limited local market, and whose costs are sensitive to their market shares. There is also the problem of successful legal action brought against a foreign producer resulting from defective products sold under its name but not produced by it. Given Thailand's prospects for moving into products and industries where proprietary technology is important, the lack of protection could prove a deterrent in the future.

There is no regulation of technology imports. Royalties, fees, and terms of agreement between licensor and licensee are left to negotiation between the parties involved.

Regulation

Complaints about government regulation refer to cumbersome customs procedures, slow decision making, ambiguous tax laws (Panyarachun & Coe 1984, 16-17), occasional retroactive changes in tax liability and customs classifications (with retroactive penalties), and intergovernment agency disputes. Thailand's first export processing zone was opened in 1983, motivated in large part by the desire to short-circuit the cumbersome process of government regulation, and in particular to expedite customs clearance. Clearance in imported shipments, which once required over 80 signatures, still requires more than 20. There is a widespread feeling that government regulation is weak and inconsistent (hence unpredictable), but not unreasonable.

These problems reflect the nature of the Thai approach to management of power: the autonomy of government branches and agencies; vague and overlapping jurisdictions; and lack of centralized authority. The Joint Public-Private Consultative Committee headed by the prime minister is working both to simplify complicated customs and port regulations and tax clearance certification that impede exports, and to speed up the processing of applications for investors' promotional privileges.

There is little complaint about the practice of side payments, which is well institutionalized, moderate, and predictable. It can be characterized as a system of "speed" or "grease" payments that allows business to circumvent the snail's pace of cumbersome agency decisions and performance.

Excluded Sectors and Occupations

The occupations and professions prohibited to aliens under the Alien Occupation Law of 1973 are primarily traditional crafts, occupations of moderate or low skill levels (with the exception of civil engineering work not requiring specialized skills), architectural work, and legal or litigation services. Exceptions can be negotiated, particularly for promoted firms. Occupational limitations on the employment of expatriates (as distinguished from numerical ceilings by firm) are flexible. In no case were these observed or reported to be a constraint.

Sectors closed to foreign firms include private utilities, rural banking, insurance, savings banks, and defense-related products. Sectors requiring Thai majority ownership and control include rice and salt farming, local sale of agricultural products, internal trade, real property, construction, and many professional and personal services. Industries where new firms (but not those established before November 1972) must have Thai majority control include printing and newspaper publishing, plywood, industrial handicrafts, fishing, several types of farming, retail trade, tour agencies, and a number of service industries. However, exemptions for closed occupations and industries are easily obtained, especially for promoted firms (Panupong 1984, 5). Foreign majority ownership and control are allowed in other industries. Joint ventures are encouraged in all industries but not required except as noted above (Mason, in Sagafinejad 1981, 447-49). The only exclusion that might affect technology transfer adversely is retailing. Some large U.S. retailers, Sears, Roebuck for instance, have provided technical assistance, credit, and an assured market to local producers when they had a significant market position in a national economy under tariff protection. They have no need to incur most of these additional costs if they have no position in the local market and are importing for the United States or other markets, as in the latter case they can choose among many suppliers in many countries.

An alien company (in which more than half the equity is foreign-owned or more than half the stockholders are foreign) may lease land for up to 30 years. A foreign company promoted by the BOI may be given the right to own land for industrial purposes.

Competitive Conditions

There is a large number of public firms in many industries. Many are badly run and losing money. Their total expenditures in 1983 exceeded those of the central government (Sricharatchnya 1983, 61-65; *Economist* 1985, 67-68). The government has been selling some of them to the private sector. However, the new heavy industrial development on the Eastern Seaboard will include public firms. In fact, the distinction between public and private firms is not at all clear either in terms of ownership or in terms of management. A new type of national corporation pooling private and public resources was initiated in 1982 in sugar and in fertilizers.

The main areas in which public companies are most hampered by poor administration and follow unrealistically low price policies are transportation, communications, and utilities, where they have a monopoly, so there are no competitors to complain. The only complaint about the competitive behavior of public firms concerns the government pharmaceutical company, which is using political means to squeeze out foreign pharmaceutical firms, or so the latter claim. All government health facilities are required to purchase only from the government firm. Private health organizations are being pressed to follow suit. However, some government health facilities continue to purchase from private drug companies. Recent increases in import taxes on a large number of drugs are viewed by foreign firms not so much as a means of encouraging domestic production at high cost in uneconomically small batches, but as a means of switching demand to the government pharmaceutical firm.

Some 40 commodities are subject to price controls. Most of them are articles of common consumption and their control is intended to stabilize the cost of living of lower income groups. Controls on rice and some other crops, however, are mainly of benefit to exporters. No complaint was received that price controls were affecting competitive positions or investment prospects adversely.

References

American Chamber of Commerce in Thailand. 1982. *This is Thailand*. Bangkok.

Ayal, Eliezer B. 1969. Value Systems and Economic Development in Japan and Thailand. In Robert O. Tilman, ed., *Man, State and Society in Contemporary Southeast Asia*. New York: Praeger.

Economist (London). 1985. White Elephant Sale. 2 March, pp. 67-68.

Eoh, Noppong. 1984. Plight of the Jobless. *ASEAN Investor*. 2 (5): 6-9.

Faith, Nicholas. 1983. Thailand: A Survey. *Euromoney* (October). i-xxxvi, 344ff.

Hanks, Lucien M. 1975. The Thai Social Order as Entourage and Circle. In G. Wm. Skinner and A. Thomas Kirsch, eds., *Change and Persistence in Thai Society*. Ithaca: Cornell University Press.

Hongladarom, Chira, and Pradit Charsombut. 1984. *Current Employment Situation with Specific Reference to Rural Unemployment*. Bangkok: Thammasat University Human Resource Institute. 31 March. Working Paper Series no. 4.

Isarangkun Na Ayuthaya, Chirayu, and Koji Taira. 1977. The Organization and Behavior of the Factory Work Force in Thailand. *Developing Economies* 15 (March): 16-36.

Kirsch, A. Thomas. 1975. Economy, Polity, and Religion in Thailand. In G. Wm. Skinner and A. Thomas Kirsch, eds., *Change and Persistence in Thai Society*. Ithaca: Cornell University Press.

Mason, R. Hal. 1981. Comment. In Tagi Sagafi-nejad, Richard W. Moxon, and Howard W. Perlmutter, eds., *Controlling International Technology Transfer: Issues, Perspectives, and Policy Implications*. New York: Pergamon Press.

Panupong, Chira. 1984. MNCs and the Role of the Thai Government. Paper presented at a Conference on the Role of Multi-National Corporations in Thailand. Organized by Thammasat University, Cholburi, Thailand, 7-9 July.

Panyarachun, Anand, and B.A. Coe. 1984. MNCs in Thailand: An Investor's Point of View. Paper presented at Conference on the Role of Multi-National Corporations in Thailand. Organized by Thammasat University, Cholburi, Thailand, 7-9 July.

Pontius, Steven K. 1983. The Communication Process of Adoption: Agriculture in Thailand. *The Journal of Developing Areas* (October): 93-118.

Santikarn, Mingsarn. 1981. *Technology Transfer: A Case Study*. Singapore: Singapore University Press.

Scott, James C. 1972. *Comparative Political Corruption*. Englewood Cliffs, N.J.: Prentice-Hall.

Sricharatchanya, Paisal. 1983. The Lame Duck Legacy. *Far Eastern Economic Review* (21 April): 61-65.

Wyatt, David K. 1975. Education and the Modernization of Thai Society. In G. Wm. Skinner and A. Thomas Kirsch, eds., *Change and Persistence in Thai Society*. Ithaca: Cornell University Press.

4

U.S. and Japanese Contributions to Technology Transfer via Human Resource Development: Indonesia

The next three chapters examine the activities of U.S. and Japanese organizations contributing toward development of human resources required for technology transfer and for increasing absorptive capacity. Chapter 4 discusses contributions to Indonesia; chapter 5 discusses contributions to Thailand; and chapter 6 compares the U.S. and Japanese contributions and also differentiates their behavior in Indonesia and Thailand.

U.S. Contributions

Organizations of U.S. origin or substantial U.S. connection that contribute to technology transfer in Indonesia have been divided into three groups: (1) private nonprofit organizations (foundations, societies, and associations); (2) for-profit business firms; and government agencies (mainly AID). These three groups differ basically in objectives and in modi operandi, and also in the nature of their contributions to technology transfer.

Private Nonprofit Organizations

It is not possible to arrive at a reliable estimate of the level of activity of U.S. private nonprofits engaged in technology transfer. The Technical Assistance Information Clearing House (TAICH) report on Indonesia of March 1982 lists a total of 67 organizations conducting development assistance programs. This list is far from complete; we have identified a number of others. Of these 67, 44 provided data on 1981 expenditures of $49.8 million. But we are warned: "These figures should be viewed more as indicators than as exact sums; differences in fiscal year, methods of financial reporting and methods of estimating dollar value of commodities, equipment and material . . . make correlation and absolute accuracy impossible." The above dollar figures do not include some donations in kind, nor do they include the value of volunteer services. Much of the expenditures listed (and transfers in kind), however, are for relief, rural and community development, with limited relation to technology transfer and associated human resource development.

Thus expenditure totals greatly overstate contributions to technology transfer and associated human resource development because of the all-inclusive nature of the totals. The totals understate their contribution to the extent that a significant number of organizations listed provide no numbers, that volunteer services and donations in kind are omitted or undervalued, and that several organizations conducting relevant activities are not included.

After eliminating some organizations whose activities were clearly not relevant, 56 were queried by mail, of which 26 responded. It is believed that the nonrespondents do not represent significant relevant activities. Six additional nonprofits, believed to be major contributors to technology transfer, were interviewed, both for information on their own activities and for their perspective on priorities and contributions of other organizations.

The contribution of those nonprofits whose activities are directly relevant is much greater than dollar expenditures alone would indicate. They have provided leadership and played a catalytic role. Because of their freedom from political pressures or stockholder demands, they can formulate coherent policies and maintain them over long periods. They also have the flexibility to respond to changing needs and opportunities that government agencies typically lack.

One may distinguish between three kinds of activities: activities contributing toward increasing the absorptive capacity of the host country in general and in specific areas; activities promoting specific technology transfer and diffusion; and activities whose contribution to technology transfer and/or absorptive capacity is not a primary mission, but either indirect and incidental, or ancillary to their major purpose.

The Ford Foundation illustrates the contribution of nonprofits to absorptive capacity. Between 1950 and 1980 it sent close to a thousand Indonesians abroad for advanced study, predominantly to the United States. Before 1965 its focus was on training public administrators, economists, and educators; since then it has contributed also to training in management, agricultural and environmental research, and regional planning. The foundation pioneered in family planning, until AID and the World Bank became interested. It is primarily concerned with building institutions, such as the Center for Demographic Studies in the University of Indonesia and the Bogor Agricultural Institute Graduate School, rather than just training individuals. In education, its focus has shifted from teacher training to educational planning, management, and research. Its grants to ongoing or completed projects in 1982 or later exceed $4 million.

The Rockefeller Foundation has concentrated its activities in the agricultural sciences and the health sciences, and in particular in building up the capability of Gadjah Mada University in these areas, as well as in the social sciences. It has financed advanced training for 190 individuals (including 96 graduate degrees), mainly overseas, and 81 man-years of senior visiting faculty, granting $1.9 million for training (which does not include the cost of visiting faculty). The Agricultural Development Council supports advanced training overseas and has sent specialists as advisers to both Gadjah Mada University and Bogor Agricultural Institute.

The International Executive Service Corps (IESC) is engaged in "retailing" managerial and technical assistance to domestic organizations, mainly business firms. The IESC draws on volunteers, primarily retired businesspeople, who typically serve three-month assignments. The advisors are not compensated and are not sent by particular firms. The fact that they have decades of experience and that they are not agents is of importance, because IESC has learned that in the majority of cases the problem for which assistance is sought is misconceived by the organization seeking assistance. Equipment and process problems in many cases turn out to be management or personnel problems instead, and the volunteer ends up as a trainer rather than technician. IESC's policy is that local organizations should pay some of the costs, with payment determined on the basis of ability to pay rather than actual costs. Since the inception of its activities in Indonesia in 1968, IESC has accepted over 400 projects in production, engineering, research and development, marketing, quality control, sales, finance, controllership, and accounting.

Volunteers in Technical Assistance (VITA) focuses on technology transfer in a wide range of areas, including technologies appropriate for small and medium-scale production. It has a worldwide network of volunteer experts and a computerized inventory of over 60,000 technical reports, which permit it to respond to requests with information and technical assistance by mail and on-site. It has also developed technologies appropriate for LDCs, such as windmills and energy-efficient stoves. It has trained Indonesians in information services for technology and skills in twice-yearly training programs conducted in Washington.

The Fund for Multinational Management Education (FMME), in collaboration with the ASEAN–U.S. Business Council, initiated a training program in 1983 on industry- and country-specific management of technology. In Indonesia it is working with the University of Indonesia's Institute of Management and the Institute of Management Development and Education (Lembaga Pendidikan Kan Pembinaan Manajemen—LPPM). Industries targeted for training programs are food processing, industries related to the petrochemical sector, pulp and paper, wood products and furniture, garments, and metal working. The training courses will be offered several times annually. FMME intends to transfer the conduct of training programs to local institutions in ASEAN nations. More recently, FMME has been instrumental in founding the U.S.–ASEAN Center for Technology Exchange. The center not only will offer training courses and technical assistance programs in each ASEAN country, but it will assist ASEAN businesspeople in learning about technology trends in the United States. The center will help in identifying U.S. sources of training, technical assistance, technology, and investment. It has initiated programs in Thailand and in Indonesia.

IESC, VITA, and FMME all receive substantial assistance from AID, as well as from a variety of sponsors including both U.S. and foreign businesses, nonprofits, and public organizations.

The third group of nonprofits, whose contribution to technology transfer via human resource development is indirect or incidental to their primary mission, is mainly concerned with improving health and living conditions in rural areas. This group provides assistance in health, nutrition, agricultural activities, small-scale industry, and related institutional development. Catholic Relief Services–U.S. Catholic Conference is by far the largest. The value of its program activities in 1983 was estimated at over $15 million. The greater part of this consisted of donations of food for distribution to the needy (Pub. L. 480, Title II, commodities from AID, milk powder from the EEC). Some of this food was donated in exchange for community development work: roads, irrigation systems, land preparation, or various agricultural and community investment projects. Other activities include leadership training; training for health-care personnel; agricultural training for new crops; a pharmaceuticals information center for medical personnel; vocational schools; development of radio stations and educational radio programs; and provision of agricultural tools. Its main agricultural development program is in East Timor.

The International Institute of Rural Reconstruction trains Indonesians for village leadership at a training center in the Philippines. Other organizations are interested in rural betterment. Some tackle specific health problems, and train practitioners to deal with them, such as the Helen Keller Foundation and the World Rehabilitation Fund.

Many of these nonprofit organizations introduce technologies and conduct training required to use them, but the technologies are in most cases simple and widely available, and the contribution is not so much to technology transfer as to adoption and diffusion. The main exception is technology-transfer and associated higher-level training in health-related areas.

What organizations concerned with rural improvement and community development contribute is leadership and example, raising awareness of problems and possibilities for solving them, supplementing local resources, modifying local attitudes, and helping to build appropriate institutions. Indirectly they contribute to technology transfer by increasing willingness to accept new ways, opening up channels of communication, and facilitating the process of diffusion.

The sources of funds available to nonprofits are diverse. The major foundations, such as Ford and Rockefeller, have their own resources, and contribute to other nonprofit organizations as well. Most nonprofits rely on contributions from sponsoring organizations: religious organizations, the Red Cross, or YMCA. Quite a few have a wide range of contributors: business firms, other nonprofits, and various public agencies. Many receive substantial funding from AID—for instance, IESC and the Asia Foundation. (AID lists 31 U.S. private voluntary organizations with which it is working in Indonesia; five of them are not in the TAICH directory, and at least 18 have received AID funds.) Some nonprofits receive funds from the international aid organizations of other nations, from international institutions (U.N., World Bank, and Asian Development Bank), and from the government of Indonesia itself. These contributions may simply enlarge the resources that the nonprofit has available to carry out its own agenda, as is the case of IESC. Or the contributions may represent some transfer of AID activities to nonprofits, as in the management of overseas training and education activities by the Asia Foundation. Many nonprofits receive no AID funds, and even those that do cannot be said to represent AID under a private nonprofit guise. They are supported essentially because AID determined that the activities they were carrying out with other funds merited additional support. None of the nonprofits referred to above are agents of specific corporations, industries, or trade associations.

U.S. Business Firms

Information on U.S. business firms in Indonesia and their contribution to technology transfer was obtained from a variety of sources. Ten firms were interviewed during the first field trip. These represented a wide range of activities in eight industries, most of them large firms. Information on additional firms was obtained during both trips. We interviewed individuals knowledgeable about an industry or the business community, including businessmen, Indonesian and U.S. government officials, and representatives of associations, educational institutions, and foundations. Published materials were surveyed. Much useful information was obtained from a report on training prepared for AID (Cox 1985). This report was based upon questionnaire returns from 74 Indonesian and foreign firms, and on follow-up interviews with firms, training institutes, government officials, and industry representatives.

Identification of U.S. firms in Indonesia is a matter of drawing a somewhat arbitrary boundary. Since 1974 only joint ventures have been allowed (although there a few exceptions, primarily producers for export). Local ownership initially is required to be 20 percent, rising to a minimum of 51 percent in 10 years. The relative decision-making role of the Indonesian partner, the local representatives of the U.S. parent firm, and the parent firm itself can vary widely, and there is no way of ascertaining this for a large number of firms. What a joint venture means in terms of U.S. management influence can range from an

Indonesian silent partner (who is simply paid to meet the letter of the law), to an Indonesian firm that is autonomous (but employs selectively the technical skills of a foreign partner and may also take advantage of the foreign partner's access to capital or markets). The domestic capital may in fact be a domestic contribution to the enterprise, or may come—suitably laundered—from the foreign investors. Majority local ownership can undermine the foreign investor's control, or have no effect on it. The American Chamber of Commerce lists members that are not U.S. firms, but that do substantial business with U.S. citizens, or happen to have a U.S. expatriate as an employee. Although the absence of clear-cut distinctions may preclude obtaining precise numbers, we easily identified U.S. firms and their behavioral characteristics.

Human Resource Development Activities. A number of firms, especially large MNEs, recruit Indonesians for managerial and professional positions on overseas, particularly U.S., college campuses. Most employers, whether or not they recruit overseas, typically seek out graduates of specific Indonesian universities or technical training programs. Much stress is placed on educational attainment. (For office and production workers, some employers in the Jakarta area, and presumably in other large cities, rely largely on newspaper advertisement and similar formal methods.)

The following types of human resource development activities facilitating technology transfer were reported by one or more of the firms interviewed and also by other respondents:

1. Training of technical and managerial new hires (almost universal), frequently in Asian regional headquarters (Singapore, Manila, or Hong Kong), less frequently in the United States.

2. Continuing training, and updating for technical and managerial personnel. This too is done by most firms interviewed, but whether it is done, and how intensively, appears to depend on the rate of change in the products or technologies of the firm. Considerable reliance is placed on instructors brought in from other firms within the company.

3. Training of customers. This is a major activity when the product is complex, and its market is a function of the ability to use it effectively, for example, IBM.

4. Training of agents and distributors. This is typically done by firms that have no domestic manufacturing facilities, but also by some firms that do manufacture locally, and whose products require servicing and maintenance. Agents may also be trained to teach customers how to use the products.

5. Training of suppliers. This may be in part a response to pressures to replace imports with domestic products; domestic production is promoted, and/or

domestic producers are assisted in improving the quality of their product, or meeting specifications.

6. Opening of in-house training programs to individuals other than employees. This is sometimes a way of screening potential employees, but in most cases it does not appear to be a response to the needs of the firm, but an exercise of corporate social responsibility or response to government exhortation. Some training programs are designed for nonemployees.

7. Financing of fellowships. The fellowship holder may be disposed to work for the firm financing the studies, but some fellowships are jointly sponsored by numerous firms, and there is no commitment by the fellowship holder to any of them.

8. Providing personnel from the firm for instructional purposes outside the firm. Again, sometimes this becomes a cost-effective way of screening for potential hires, but there is no commitment.

9. Providing equipment, supplies, facilities for educational and training institutions.

10. Financial support to educational and training institutions, rather than to trainees or students.

All U.S. firms interviewed in Indonesia conducted training activities for their new employees. Although some had large continuing efforts, others had terminated their regular training programs because all their employees had adequate training. Other reasons cited were very low turnover, little change in employment, and hence minimal need for new hires. One firm terminating its program relied on another U.S. firm that conducted training programs for workers other than its own employees; the former firm's own requirements had become too small to justify an in-house training program. (One other characteristic of firms whose training efforts had been terminated for reasons other than low turnover and no expansion may be involvement with a static or relatively static technology.) Goodyear Indonesia is an exception to the rule that deserves special mention, although its training is for production workers rather than managerial and professional workers. It provides high-quality training for outsiders, which is highly regarded, and has served as an example followed by a number of other foreign firms at the urging of the Department of Industry. Goodyear initiated extensive training programs for its employees in the early 1970s, including a large number of self-study modules available in Indonesian. It opened an Apprentice School in 1973 with a three-year program of instruction. When a few years later it no longer needed most of the trainees, it decided to continue the school, with graduates free to seek other employment. All graduates have been placed, and some have moved up to engineering and training management positions.

Signals are mixed as to the responsiveness of training to labor market conditions in Indonesia. Most firms assert that their training is exactly what they do elsewhere. The frequency of regional training programs, with new hires sent out of the country for part of their training, is in accordance with their assertion. Regional training implies that training is not remedial and is not adjusted to country conditions. On the other hand, the same firms that train new managerial and professional hires on a regional or MNE-wide basis, may hire predominantly outside of the country, or give strong preference to hires with university education in the United States and other advanced countries. They may circumvent the need for remedial training (or for adjusting training to country needs) by not hiring Indonesians trained exclusively in domestic institutions. Regional headquarters training requires knowledge of English, which students in the United States possess.

All firms interviewed, with one exception, claim that nearly all new hires successfully complete their training programs. The one exception is a very high-technology firm, the majority of whose employees are expatriates.

Turnover and Training. Turnover of trained, experienced managerial and technical personnel from foreign firms to domestic firms is a major process of technology transfer. It induces additional training. Low turnover can be the result of a number of circumstances. One is that the technology-related skills are in industry-specific technology. This can be found in technical and professional areas, whereas managerial skills are less likely to be industry-specific, particularly in the financial and accounting fields. Turnover was reported to be high in services relative to goods-producing industries. An industry-specific skill (as in petroleum) can still mean high turnover if there are a number of firms in the industry in the country; thus the number of firms is important. Another reason for low turnover is government policy. Pertamina is said to frown upon interfirm raiding in the petroleum industry. This is the reason one firm in this industry finds it unnecessary to maintain a training program for new entries. A further reason is considerable disparity in pay, fringe benefits, and working conditions between firms in the foreign sector and those in the domestic sector of the same industry.

Although managerial and professional turnover is generally reported to be very low, we found it to be high in two sectors: in banking and finance, and in construction. The nature of construction employment requires occasional changes of employer; much of the turnover is involuntary, a response to phasing in and out of projects. In the case of banking and finance, however, the turnover is almost entirely voluntary, and very high, in some cases from 30 to 50 percent.

In the banking industry, domestic banks are able to pay an experienced manager rates above those prevailing in the foreign sector. Partial deregulation has increased the need for such individuals, and their experience also proves attractive to firms and government organizations outside banking. (One respondent, who hires exclusively in Indonesia, attributes the high turnover in the financial sector to the fact that employers scour foreign campuses for those candidates who are most in demand and therefore most mobile. But the same practice does not lead to high turnover rates in other sectors.) To a considerable extent the high turnover (to foreign firms as well as to domestic organizations) and associated large training programs compensate for the inadequate supply of well-trained graduates of local institutions.

Pertamina's discouragement of turnover (raiding) between foreign-sector firms has little effect on technology transfer from the foreign to the domestic sector. But other government agencies are reported to encourage unproductively high turnover, mostly between foreign firms, apparently with the expectation that this will compel the firms to increase training efforts to the benefit of the Indonesian economy, without regard to what these efforts surely do to the costs of production. The stop-and-go nature of project-related activities in the construction contracting area, the insistence that a new organization be created for each new project, and short deadlines compel firms obtaining contracts to raid other firms as the only way to meet deadlines—there is no time for training. Thus raiding substitutes for training among some firms, even while the training needs are increased in firms suffering from raids.

It is not clear that much additional training is done, but it is clear that costs go up, not only because of the costs of successful raiding to the raider, but because of the disruptive effect of raids on the productivity of the victimized firms. These effects, combined with repeated formation of new organizations to implement new projects, assures that the industry always remains low on the learning curve. The additional training that occurs is at lower levels of skill where extensive raiding is not possible (or is prohibitively expensive compared to the costs of training). But this is largely the result of changing geographical location of large construction projects as well as their discontinuous nature, not the result of policies designed to maximize the training contribution of firms. At the same time that surplus construction workers are sent abroad, there are complaints of shortages of construction skills at home (and of the inability of Indonesian firms to manage large projects overseas). These two situations coexist both because of geographical differences between the location of skilled construction workers and of demands for them, and of policies favoring employment of local workers, therefore training them, over relocation.

The high voluntary turnover in and from the finance industry certainly does not support the common view that Indonesians are reluctant to leave an employer when attractive alternatives exist.

Even where much of the turnover is among foreign-related firms, such turnover indicates a functioning labor market that allows movement to domestic firms, and a supply of experienced workers willing to change jobs. Thus the absence of turnover in many industries must be attributed to a deficiency on the demand side. The confinement of turnover largely to foreign-related firms indicates either a lack of demand among domestic firms or an unwillingness or inability to offer conditions of employment competitive with foreign-related firms.

Status of Training among U.S. firms in Indonesia.   It appears that in the last several years net U.S. investment in Indonesia (excluding the oil and gas sector) has been zero or negative. This conclusion was suggested by the absence of a considerable number of firms from the latest *Directory of American Business* in Indonesia that had been in the previous directory. Knowledgeable individuals have confirmed the withdrawal of a substantial number of firms and have also expressed the opinion that net investment had been zero or negative.

Under these circumstances there is precious little technology transfer by U.S. firms via human resource development. Training programs that were active when the firms were first established and that were maintained as they expanded their investment, have since been terminated. Net new investment is one major indicator of the need for training. The other is turnover. Even without growth, a firm must continue training to the extent that it can expect to lose experienced people (or hire laterally to replace them—something all firms cannot do when there is a shortage of experienced managerial and professional workers). But turnover is low. Firms of other nationalities are also holding down investment, and the domestic economy is depressed. New firms are not being formed nor are existing firms being modernized to attract experienced workers from foreign firms.

The sector that continues to do a great deal of training—the banking and finance sector—is precisely the one with high turnover. This is partially a result of growth and modernization of domestic financial institutions.

Many firms—including banks—train their managerial and professional employees in regional centers rather than in each country in which they have an affiliate. These firms have enough new hires and turnover replacement needs to require and justify regional training programs. But the number of trainees from Indonesia fluctuates with the level of new hires in Indonesia.

Firms working largely on a project basis (engineering consulting firms and construction contractors), experience wide fluctuations in employment within project cycles and with the phasing in and phasing out of specific projects. Because projects are located in different regions, and often in remote areas, recruitment and training of workers in the project area accompanies every project. Thus we see simultaneous export of surplus construction workers to the Middle East and large-scale training programs for the same kinds of workers at home. Oil and gas companies are in a special position. Their continued investment, and the temporary nature of some of their activities often in remote locations, present continuing needs for training. For production-sharing producers, training is extremely cheap, since Pertamina bears 89 percent of the reduction in profits resulting from training costs. A consortium of companies, predominantly U.S. firms, finances advanced training at U.S. universities from company funds with 110 Indonesians currently abroad (1985).

There are a few other exceptions—firms in technologically dynamic areas—whose managers and professionals must undergo frequent updating, even in the absence of new investment or turnover. But this is a training activity different from that of new hires, and is primarily undertaken by longer-term employees, that is, workers associated with low turnover rates and therefore with correspondingly low technology transfer.

What applies for training by firms for their own employees applies also to training for suppliers. Firms are not increasing their demand, and are not anxious to subcontract more of their work or to invest in developing local suppliers to replace imports in a slack period. Where customer training is an integral part of marketing and postsales service, it is less affected by the current slowdown and pessimistic expectations. As to contributions to human resource development beyond the needs of the firm, its suppliers, and to its customers, it would be surprising if these contributions were to remain unchanged. If firms are not hiring, and their profits are down, one would not expect them to have as much incentive, or as many resources, for contribution of funds, equipment, and instructors to local training and educational institutions, or to scholarship funds and the like. There is no evidence on trends in such corporate contributions, but this is the logic of the present situation.

With little hiring and little turnover, firms are presumably moving up on the learning curve, through a "learning-by-doing" process. As to systematic, formal training programs, however, the impression is that many firms, if not most firms, have discontinued them.

The conclusion is that two factors are the key determinants of U.S. related contribution to human resource development and associated technology transfer. One of these is net new investment, in expanded capacity and especially in new products and processes. The other is a high demand for trained management and professional employees that will induce significant turnover, and/or a business climate that will induce some of them to start their own firms. This high demand must originate in the domestic sector. High demand from other foreign-related firms will induce circulation of workers among them, with a redistribution of training efforts and costs (on a net basis) from firms hiring to firms experiencing a significant turnover. Turnover of entrepreneurial individuals and trained professionals to the domestic sector will in turn induce further turnover, but on the whole turnover is dependent on high demand; it is not autonomous.

Corporate Structure and Organization of Training. The true multinational corporations are often organized in regional divisions. Typically much of the training for new managerial and professional employees in such companies is conducted in their regional headquarters. This is a disadvantage for countries such as Indonesia that are not regional headquarters countries. New hires, as an extended part of their training, may be sent on short tours of duty to firms in other countries. This is of less benefit to the countries from which they are sent than to countries that "export" this on-the-job training, on a net basis. U.S. MNEs, for a variety of reasons, are more likely to follow this organizational structure than MNEs of other countries. On the other hand, they are less likely to send new hires for training to the world headquarters country than Japanese MNEs. Training conducted at home, where feasible, is of greater total benefit to the country than training conducted elsewhere.

For firms that are not organized regionally, the focus of training for U.S. firms is predominantly in-country. Company headquarters in the United States are far away, and training of new hires there is expensive. Furthermore, U.S. companies are likely to delegate extensively to their foreign affiliates, including delegation of training. A few companies routinely send new managerial and professional hires to the United States, but most companies, if they send employees to the United States at all, do so only at a later stage in the careers of new hires, and do so selectively. They send employees expected to advance to more senior positions. To the extent that training capability is not built up in the LDC, the benefit to the LDC is reduced. On the other hand, the workers are likely to receive better training.

The regionally organized firms tend to have more formal and structured training programs, because the programs accommodate trainees from diverse countries. With this kind of training and overseas experience, the managers and professionals are likely to be more mobile than those whose training is very much on-the-job, or specific to an establishment or corporation, particularly when the establishment has a great deal of autonomy. Thus regional training is likely to lead to greater turnover, hence to greater technology transfer.

Employment of Expatriates. The firms we interviewed varied widely in their use of expatriates. One firm employed only two expatriates; another reported that 50 percent of its employees were foreigners. This difference reflects the kind of technology the firms deal with, not the age or size of the firm. In fact, the number and proportion of expatriates early in the operation of a foreign firm in Indonesia is a pretty good indicator of the technology transfer potential of its activities. The reduction in the number of expatriates during the course of a firm's experience in Indonesia is an indication of the extent to which their replacement by adequately trained and experienced Indonesians (the first step in technology transfer) has been accomplished. Firms interviewed indicate that five to six years is the time required to replace expatriates except at the most senior level. This is a period in keeping with the Indonesian government's expectations for manufacturing, but longer than its expectations for services. Although a major function of expatriates is the training of their replacements, not all expatriates are to be viewed as temporary or as trainers; some remain as representatives of the parent firm to exercise management control.

Attitudes varied toward Indonesian government policies and the pressures to reduce the number of expatriates. There were complaints, but they were not universal, and no respondent claimed large effects on his own firm's cost of production, or its ability to function effectively. Assertions of severe impact generally came from Japanese firms. Obviously the effect is felt most by firms that have only recently been established in Indonesia. These firms, unlike long-established firms, have not had the time to develop their own Indonesian technical and managerial personnel. However, outside the petroleum exploration and development sector, the proportion of expatriates is low whatever the age of the firm. This suggests that the products and production processes involved are not high-technology, that the technology transfer potential is therefore limited.

It is not clear that Indonesian government policies have significantly accelerated the replacement of expatriates by Indonesians, that is, they have not induced expanded training and human resource development activities by the firms. It is unclear whether would-be investors' perceptions of Indonesian government restrictions on the use of expatriates have been a deterrent to U.S. investment, in particular for higher-technology products and processes that require a substantial proportion of expatriates for a number of years. The cost of expatriates may be deterrent enough. Probably there was no deterrent effect before the highly publicized tightening of regulation and enforcement in 1984, which could bring adverse reactions. One firm reported that its growth was held back by restrictions on the number of work permits for expatriates.

Domestic Content. The question of import substitution policies and pressure as they affect U.S. firms was not explored systematically. There were no reports on training for suppliers that linked them to domestic content requirements. (However, the pharmaceutical industry is reported to be under pressure.) Most of the reports on pressures for domestic production of parts and components otherwise more cheaply imported were heard from Japanese manufacturers. Knowledgeable respondents felt that policy on replacing imports with domestically produced products is pushed to the point of serious counterproductivity for the Indonesian economy. Not only are the costs of the domestic product very high compared to imports, but also the possibility that widely known pressure to replace low-cost imports with much higher-cost domestic production deters entry of new foreign firms. These firms' products or productive processes might be vulnerable to pressure; such pressure certainly precludes development of export markets. There is independent evidence that import substitution has been pushed too far (Gray 1982, 43–48). The counterpart of import substitution—restrictions on export of unprocessed materials, specifically logs—is reputed to have been a factor in the withdrawal of U.S. firms as an alternative to investing in processing facilities. Thus it is not clear whether the effect on the amount of technical and managerial training, and on the accompanying technology transfer, is positive or negative. This assessment of current practices does not imply that a domestic content policy cannot be beneficial by spurring laggard firms toward import substitution.

U.S. Investment. The scope and nature of the contribution by U.S. business to technology transfer via human resource development is best indicated by the volume and industrial distribution of U.S. investment in Indonesia. There are two sources of information. The first is based on investment approvals by the BKPM. Not all approved investment plans are implemented, and to the extent that they are, there are lags; these estimates are therefore approximate. Unfortunately, there is no definitive information. The estimate for cumulative investment is roughly $600 million, and has been for years. This total excludes investment in the oil and gas industry, estimated at $4 billion between 1971 and 1980 (Wie 1984b, 92), as foreign firms are not permitted ownership positions in this sector. All oil and exploration and production companies except Caltex are operating on a production-sharing basis. Estimates for recent annual investment of U.S. firms in this sector run as high as $2 billion per year. This figure excludes investment in banking and insurance. Cumulative manufacturing investment approved is estimated at around $300 million.

The other source is from the *Survey of Current Business* (1985, 35–40), which reports annually the U.S. direct investment position abroad. At the end of 1984 the total was $4,409 million, of which the petroleum sector accounted for $3,892 million and all the rest for $617 million. This source reported U.S. investment in manufacturing at $152 million, a figure that has changed little for the past five years.

The estimates based on BKPM approvals give a better picture of total U.S. investment in Indonesia, which is the pertinent indicator of technology transfer potential. The *Survey of Current Business* figures reflect capital outflows from the United States, which are smaller, as not all investment by U.S. firms in Indonesia is financed by capital outflows from the United States. What is the implication of the fact that capital flows from the United States to Indonesia are smaller than the investment by U.S. firms in Indonesia? In terms of human resources development, it is the amount of investment, not its country of origin, that matters. But in terms of the U.S. contribution to human resource development, one might claim that only capital flowing in from abroad and its associated human resource development should be counted. Capital originating in Indonesia may simply displace other investments and their associated human resource development contribution. This counsel cannot be followed because of lack of information as to the ultimate origin of U.S. investment funds, whether these be the United States or other domestic and foreign sources. But in the case of Indonesia, savings have exceeded investments for much of the past decade. Therefore, Indonesian savings invested in U.S.-related firms do not seem to represent an opportunity cost—investment foregone in other firms.

Apart from oil and gas, the U.S. investment is concentrated in manufacturing, in chemicals, rubber, pharmaceuticals, electrical machinery, and semiconductors; and, in services, investment is substantial in trade, banking, and finance.

U.S. industry in Indonesia, oil and minerals aside, is notable mainly for its absence. Cumulative investments that total a mere $600 million are ludicrously small in relation in Indonesia's size and potential, and to U.S. investments in other smaller Asian nations. Except for two manufacturers of computer chips solely for export, U.S. investments almost exclusively serve the domestic market. Others find foreign investment in manufacturing to be surprisingly small (McCawley 1979, 66–67).

Why is U.S. investment in Indonesia so limited (oil and gas excepted)? Why is the investment predominantly for domestic markets and not for export? (It is not surprising that the investment is not high-technology and that it does not involve a great deal of technology transfer, given the shortage of highly skilled and technical labor and the shortcomings of educational and training institutions.) But why has investment stagnated?

A number of interviewees volunteered the information that U.S. firms well established in Indonesia were doing well. (See also Wie 1984a, 100.) This view appears to be somewhat at variance with the withdrawal of a number of U.S. firms from Indonesia. When one looks at the small cumulative investment and particularly the near-disappearance of net investment since the mid 1970s, one must conclude that the counsel of U.S. businessmen in Indonesia to prospective investors must have been discouraging. The Indonesian government is concerned about the lack of U.S. investment, for it has taken measures to promote joint ventures specifically between U.S. and Indonesian firms.

Whatever OPEC may have meant for the rest of the world from 1974 on, for Indonesia (and other major oil exporters) it meant a great increase in resources and opportunities for development and growth that did not exist before. The great increase in foreign exchange earnings for the foreseeable future had to make it a much more attractive place in which to invest. However, the number of new U.S. investments in Indonesia dropped sharply from an average of 11 a year during the late 1960s and early 1970s to two-and-a-half from 1974 to 1982 (U.S. Embassy 1983, 15–17). Granted that much of the investment in the earlier period was reinvestment by U.S. firms already in Indonesia that responded to the improved political and economic climate after 1967, the attractiveness of Indonesia still appears to have declined.

There are three possible interpretations of the abrupt decline in new investments after 1974. The riots in early 1974, mainly nationalistic and anti-Japanese, could hardly have been encouraging to other foreign nationals. The decade of stability since then should have eliminated this factor as a deterrent to new investment. Because the lack of new investment persists, one cannot escape the conclusion that the 1974 amendments to the Capital Investment Law of 1969 were the main factor. The amendments required joint ventures with Indonesian partners with a minimum 20 percent Indonesian share and Indonesian majority ownership within 10 years, set limits on the number of expatriates, and made provision for their phase-out (Donges et al. 1980, 394–5). These legal requirements are still on the books (1986) and are being more tightly enforced.

A third explanation offered for the lack of U.S. investment in Indonesia (and in particular the lack of production for export) is the overvaluation of the Indonesian rupiah since the large increases in oil prices in late 1973 and 1979. The reasoning is that large abrupt increases in dollar earnings keep up or push up the value of the rupiah, whereas large increases in domestic spending bring about accelerated inflation. As a result, Indonesia becomes a high-cost country in which to produce for export, and both domestic inflation and overvaluation of the rupiah raise costs of production for import-competing industries.

This hypothesis was tested using 1972 exchange rates and prices as a starting point. (It matters little whether this year, or the period 1970-1972, is used as a baseline.) Then the 1978 exchange rate was compared with the exchange rate that would have obtained had the rate reflected the same relation between prices in the United States and in Indonesia that prevailed in 1972. We used the GDP deflator for Indonesia and the GNP deflator for the United States. The same procedure was used to calculate exchange rates for 1981, using 1972 and 1978 as the baselines. (It is not assumed that 1972 represented a purchasing power parity equilibrium exchange rate; all we are doing is estimating the divergence from that rate associated with the large increase in oil prices in 1973 and again in 1979.)

	1972	*1978*	*1981*
Actual exchange rate, rupiah/ dollar	415	442.05	631.76
Calculated exchange rate, with 1972 as base year	415	872.05	1,260.43
Calculated exchange rate, with 1978 as base year		442.05	638.93
Ratio of calculated to actual exchange rate with 1972 as base year		1.97	1.995

Thus the rupiah had nearly twice the exchange value in 1978, and also in 1981, relative to 1972, that it would otherwise have had on the basis of trends in domestic prices in Indonesia and the United States. It was the 1973 oil price increases that led to "overvaluation"; the 1979 increases had no further effect on the exchange value of the rupiah. In 1983 the rupiah was further devalued to 970 to the dollar. In recent years the dollar admittedly has become overvalued, so that it is very difficult to judge whether the rupiah is still overvalued with respect to the relation between Indonesia and United States domestic prices prevailing in 1972.

That the "overvaluation" of the rupiah should have some negative effect on U.S. and foreign investment generally (aside from oil and gas) cannot be questioned, but it is doubtful whether this is a major explanation for the decline in new investment. Overvaluation was a gradual process that took years, whereas the drop in new investment was more abrupt and came early in the process of overvaluation. It is not plausible to assume that investment decisions were made with perfect foresight, or even in anticipation of OPEC impacts on exchange values.

The views of experienced Indonesian and U.S. businessmen and of other knowledgeable individuals were obtained on factors deterring U.S. investment. One kind of explanation concentrates on the characteristics of U.S. businesses. These include U.S. businesses' traditional dependence on a huge domestic market; the lack of the equivalent (until very recently) of the Japanese export trading companies that would handle exports and eventual overseas investments for smaller firms lacking the knowledge or resources to do it on their own; the high rate-of-return requirements; the short payback period requirements; and the large minimum market sizes for entry.

Another kind of explanation stresses local factors (particularly regulation) that translate into high costs of business entry and of doing business, and pervasive uncertainty of the local business environment— all of which deter investment. A number of the businessmen interviewed stressed this second set of factors, as one would expect them to. Several also stressed U.S. regulation, in particular the Foreign Corrupt Practices Act. But a minority stressed the characteristics of U.S. business, and one noted that the behavior of U.S. business was no different before the passage of the Foreign Corrupt Practices Act than after.

Clearly these two sets of explanations are not mutually exclusive. Characteristics of U.S. firms cannot be alone responsible, once one observes high levels of U.S. investments in other Asian countries. A comparison of U.S. firms with Japanese firms should help to distinguish between the Indonesian environmental factors, which would be approximately the same for Japanese and U.S. firms, and the U.S. business factors.

The lack of competent management was perhaps mentioned more often than any other problem. It was mentioned not only with reference to business firms, but in regard to government administration and regulation. The qualitative shortcomings of educational and training institutions for preparing skilled, technical, professional, and managerial workers were also stressed. Other problems were mentioned by respondents without any prompting by the interviewer. Some mentioned difficulties with the physical infrastructure, particularly transportation. One complained about running transportation services for his company for lack of an adequate public alternative. (It should be noted that the large increase in foreign exchange earnings from oil in the 1970s led to an investment boom by government and, to a lesser extent, by domestic private firms. This boom raised demand for managerial and professional and technical employees, as well as for infrastructure, thus aggravating existing shortages.)

There were some complaints about inadequate protection for intellectual and industrial property rights of foreign companies. Counterfeiting, patent, and trademark infringement were problems. In fact, Indonesia has no patent law. It has trademark legislation, but enforcement is not adequate. In part for these reasons, licensing of foreign technology to domestic producers appears rare.

The nearly universal complaints about government regulation do not refer to the regulations themselves, most of which are regarded as reasonable, but to their implementation or lack thereof. Information on regulations is often hard to obtain; regulations are changed without notice, occasionally retroactively. The process of obtaining necessary licenses and permits for investment, despite government efforts, remains protracted, complex, expensive, and uncertain, and is subject to a poorly organized system of side payments.

Certain sectors are barred to foreign investment: oil, strategic minerals (except for output-sharing agreements), transportation, communications, defense industries, and atomic energy. Others, including banking, insurance, construction, and legal and accounting services, are restricted. These restrictions undoubtedly reduce the U.S. investment, although they or similar ones are found in nearly all LDCs. The new countertrade policy is not favorably viewed by respondents who volunteered comments on it, although it is not aimed at MNEs, and only affects those firms with government contracts.

Some problem areas were not mentioned by any respondent, and presumably are not problems. Government firms are to be found in many sectors where private firms are also operating: steel, cement, sugar, and flour milling. The government also sets prices for a number of key commodities. The one firm asked explicitly about the effects of government competition and/or price setting on its activities did not feel adversely affected. The fact that no complaints were volunteered supports the respondent's view that government competition is fair and the prices set by government are at a competitive level, without adverse effect on MNEs' ability to compete. Nevertheless, if there were no government productive capacity and market share, it is reasonable to believe that there would be more U.S. investment.

Licensing and Franchises.   Most of the licensing by U.S. firms is believed to be with joint venture partners, not with independent firms. Thus looking at both investment and licensing activities can involve substantial double-counting of the latter. Franchises that involve trademarks and goodwill rarely involve much technology transfer (and that mainly managerial) and tend to be with independent local firms, not joint ventures. There is no evidence that license payments by joint venture partners reflect technology transfer that would not have occurred otherwise, or that they are meaningful indicators of the amount of technology transferred. Ceilings on license fees and limits on the duration of agreements invalidate them as indicators of technology transfer, and restrict licensing as a method of technology transfer to independent domestic firms. Separate agreements for technical assistance and management contracts have been used to circumvent limits on receipts through license fees. The lack of a patent law (which has been in the wings for many years now) is also a brake on technology transfer. The information available is that little licensing has been done; most of it has been done in the pharmaceutical industry, which faces a ban on imports of required ingredients. Given Indonesia's stage of development, the contribution of licensing to technology transfer would be modest at best, in the absence of any impediments.

There are two sources of information, both from the Bureau of Economic Analysis of the U.S. Department of Commerce. One is a benchmark census based on detailed questionnaires and a 25 percent sample for 1977 (1982 data are not yet available). The census reports $3 million paid in fees and royalties by Indonesian affiliates of U.S. parents, plus $15 million in service charges and rentals. A good part of this latter sum could be management and technical assistance payments associated with technology transfer. Industry breakdowns are limited in the interests of confidentiality. Chemicals appears to be the major industry involved. These data understate technology payments in two ways. They exclude payments from nonaffiliated firms to U.S. firms with affiliates in Indonesia, and they exclude payments to U.S. firms that have no affiliates in Indonesia. However, most payments for technology are by affiliates; payments by nonaffiliates are predominantly for trademarks and franchises, involving little technology transfer.

The other source is based on an annual survey that is much less detailed. The grand totals, which include royalties, license fees, service charges and rentals for tangible property, and film and TV tape rentals, are large and meaningless figures:

Year	Total Technology Payments	Technology Payments Excluding Petroleum
1978	$44 million	$16 million
1979	$45 million	$12 million
1980	$58 million	$21 million
1981	$86 million	$39 million
1982	$61 million	$43 million
1983	$63 million	$41 million

The last three reflect a very large increase in unspecified service industries and banking. A more meaningful figure for technology payments is royalties and license fees only, which rose from $3 million in the late 1970s to $6 million in 1983. Nearly all of this comes from manufacturing, and within manufacturing, chemicals account for the bulk of payments, followed by electric machinery (much industry detail is suppressed for reasons of confidentiality).

Contractors and Consultants. An important contribution to technology transfer by U.S. firms must be mentioned. This contribution is scarcely reflected in U.S. investment. This is the contribution of contracting and engineering consulting firms, a contribution associated with investment, largely by the Indonesian government and public firms, rather than by foreign firms. These are the only for-profit firms that make technology transfer their business. Contractors managing major construction projects (oil field and mining development, pipelines, processing plants, refineries, and manufacturing plants) train large numbers of workers in the skills required for the construction phase. Most of the skills do not constitute technology transfer but do enlarge the skill base, thus facilitating technology transfer. However, they also provide training and experience for Indonesian engineers and other technical and managerial employees, transferring some investment as well as operational capability. In addition, the contractors typically also train workers to operate and maintain the facilities once completed, and this constitutes partial transfer of technology not available domestically, although only operating technology.

Engineering consulting firms also provide technical assistance in improving existing productive facilities and in formulating and implementing regional development programs and projects. This is an activity that is primarily high-level training, consisting of technology transfer beyond the mere operational level. It encompasses investment and innovation capabilities as well. Not all of this is adequately reflected, or reflected at all, in investment figures, and little of it in investment by U.S. organizations. It should be emphasized that all the technology involved is open, and the only thing the contractors monopolize is know-how. As in the case of other firms, the number of expatriates may indicate the potential for technology transfer, and the extent to which they are replaced by trained and experienced Indonesians may indicate the completion of the first stage of the transfer process.

Government Agencies

Most of the U.S. government contribution to technology transfer is routed through AID. AID, however, is a performer only to a limited extent. It is mainly a funding agency and clearinghouse for information, decision making, and contracting. The actual technology transfer may be through other government agencies, for example, the Department of Agriculture, or the universities. Universities have not been included under private nonprofit organizations (many of course are state universities) because their technology transfer activities are not autonomous but are a function they perform under contract with AID and other sponsoring organizations. Project aid may be implemented by business firms as contractors.

AID provides substantial amounts of development assistance in the form of grants. But the larger part is in the form of loans. Totals tell us little, and do not permit comparisons with other organizations. Nor do totals allow comparisons over time, unless loans are converted into a grant equivalent, taking into account not only the extent of subsidy in the below-market interest rates, but also the extended repayment periods and initial grace periods. AID, as mentioned before, has a diversity of objectives, only some of which are related to economic development. Not all economic development programs and projects have technology transfer and associated human resource development as a major activity. Fortunately AID in Jakarta made a survey of training activities conducted or supported by all major foreign donors in 1982, on the basis of which it is possible to arrive at very rough expenditure estimates (AID/Indonesia 1983).

AID's contributions to private and voluntary organizations (PVOs) development programs for the fiscal years 1974 through 1984 totaled $19.7 million, of which $15.3 million had been disbursed by the end of June 1983 (AID/VHP 1983). However, all of the projects reported began in 1976 or later, and most began in the 1980s. By far the largest is the East Timor Agricultural Development Program, started in 1981, and conducted by Catholic Relief Services, to which AID is contributing $4 million, just over half the project cost. These totals refer only to contributions to projects in Indonesia. In addition, AID had granted, as of the end of June 1983, a total of $61.2 million in centrally funded matching grants to PVOs with activities in Indonesia as well as in other countries. It is not possible to break down this grant amount by country. Among the 14 recipient organizations are VITA, IESC, and Appropriate Technology International (ATI). IESC received $23 million for the period 1978-83. None of the other centrally funded grants began earlier than 1978, and only four began before 1980.

The bulk of the projects are for agricultural-rural-community development, including some small infrastructure projects: transportation, sanitation, water supply, and land preparation. There is extensive training, but the projects are primarily marked by technology diffusion rather than by transfer. Some of the training, in fields of health, agriculture, education, and management, is higher-level and does involve a substantial amount of technical assistance and technology transfer. There is also some investment in the production of books and other written materials, and in radio programs and communication.

The General Participant Training Program is a major AID contribution to increasing Indonesia's absorptive capacity. Program II, starting in 1982 and to be ended in 1987, includes loans of $9.5 million and grants of $2.6 million. It sends

Indonesians abroad, primarily to the United States, for advanced education and training in areas of priority for development, but not as part of specific AID projects. This is a continuation of a program that sent over 8,000 Indonesians for advanced education abroad, predominantly but not exclusively in the United States.

In addition to the above-listed activities, AID had projects in progress as of 1983 totaling $64.4 million in loans and $14.6 million in grants; the projects consisted mainly of educational and training activities. (An additional $10.5 million was reported for projects in higher education and agricultural education development completed in 1982.) These projects range from three to six years in duration. They include management development ($4 million); professional resources development for Ministries of Finance, Education, and Agriculture ($10.9 million); agricultural education ($22.6 million); education development ($9 million); local government training ($9.5 million); health ($11.5 million); and rural-agricultural development ($11.1 million). Nearly all include overseas graduate education (mainly in the United States) as a major part of their activities, and many also include technical assistance.

Other projects for which no dollar figures were given (in the areas of agriculture, rural development, and family planning services) provided overseas training for 77 doctoral students, 255 masters students, 127 trainees in shorter programs in the United States over 2,000 in-country trainees, and over 18,000 in-country trainees through extension services. Still other projects—training individuals to run and to function in key organizations related to development, and to staff and upgrade high-level education, research, and training facilities— provided neither numbers nor dollar amounts.

Most AID projects involving training are not technology-specific, but enhance Indonesia's ability to absorb new technologies. The majority of projects are related to agricultural and rural development, strengthening agricultural programs in universities, fostering agricultural research, supporting agricultural research institutes, and training leaders for agricultural and rural development activities. Much of the training is for government workers, to improve the planning, programming, and administrative capability of agencies involved in various aspects of economic and social development; and for university faculty, to improve their teaching and/or research capabilities. Most of the projects involve graduate education at the master's or doctorate level, much of it in the United States and other countries. Because the government (in addition to normal administrative and planning functions) does operate many public enterprises, some of the training of government employees will contribute directly to technology transfer.

Very few AID projects provide technology-specific training for operational purposes. Those that do (rural electrification, transmission and distribution; irrigation and rural public works projects) contribute more perhaps to diffusion of technology already available in Indonesia than to introduction of new technology.

Recently AID has been moving toward direct technology transfer through greater stress on the role of private enterprise. It is supporting some private nonprofit organizations that are involved in technology transfer, such as the IESC. And it is giving much attention to management training, which only recently has become a separate AID function. It now has management consultants in at least eight ministries and is contributing loans for mid-career management training and for business school research and curriculum development. Together with the World Bank it is considering the possibility of a doctoral program in management.

For the rest of the decade it sees its main priority as contributing to the creation of rural off-farm employment opportunities for the rural population. AID envisions management training for smaller business, overseas training to strengthen financial institutions, and experimentation with institutionalizing rural credit and savings programs as ways of implementing this objective.

The Department of Defense conducts training programs for the Indonesian military; most of the programs are directly applicable to civilian uses. They involve mainly training in the operation and maintenance of communications and transportation equipment, although some training in management is also given. For fiscal year 1985 the planned budget is $2.7 million, and the number of students, 250. The level of activity has fluctuated widely in the past, primarily as a function of equipment procurement by the Indonesian military.

Various other government agencies contribute to a limited extent through training programs with diverse funding sources. These programs are not aimed at any particular LDC, but citizens of Indonesia have been among the participants. The Department of Commerce has training programs in national income accounts and its Census Bureau has one in applied statistics. The Department of Agriculture's Office of International Training and Development is another example. At the request of the Indonesian government, the Department of Interior's Bureau of Mines held a conference in Indonesia on mining technology in October 1983.

Overseas Private Investment Corporation (OPIC) encourages private investment in LDCs primarily through insurance, loans, and loan guarantees, particularly for smaller businesses. It subsidizes feasibility studies, organizes investment missions, and has a computer data system for matching investor interests with overseas opportunities. Of direct relevance to this book, OPIC supports training and education of LDC nationals involved in OPIC–supported projects as well as programs by private organizations for the transfer of technology. It is contributing to a program of management training seminars in ASEAN countries, the first of which, in food processing, was held in Manila in 1983. This program is country- and industry-specific. The program is also sponsored by the

ASEAN–U.S. Business Council and the U.S. Chamber of Commerce, and is run by the FMME.

Contributions of Different Agents Compared

Not surprisingly, the type of human resource development done by business firms is quite different from that characterizing AID's projects. Firms are providing fairly specific training for their own needs; AID is funding general and professional education for national development. It is true that AID in the past, when its program was more project-oriented, provided for more technology-specific and on-the-job training as part of completing the project and creating the ability to operate and maintain it. And it is true that business firms to a limited extent support general education. Nevertheless, the differences are sharp. Nonprofits such as the Ford and Rockefeller Foundations have been doing much the same thing as AID. They provide graduate education in selected fields, and build educational and research institutions. Many nonprofits provide specific skills on a small scale, from rural homemaking skills to the technical assistance of IESC, whose activities resemble those of business firms, to which they are directed.

The difference in sectoral emphasis is at least as sharp. AID has placed and continues to place great stress on agriculture, rural development, and health improvement. U.S. firms, on the other hand, are heavily involved in oil and gas exploration and development, insurance and banking, pharmaceuticals, and construction. Most of the nonprofits doing anything remotely related to technology transfer, in particular the Rockefeller and Ford Foundations, and the Agricultural Development Council, emphasize the same sectors as AID. The contributions of AID and nonprofits, and of business firms, different though they may be, are complementary in the long run.

Japanese Contributions

The Japanese contribution to technology transfer in Indonesia is limited to the activities of business firms and government agencies. There is no large, long-established, and diverse nonprofit sector comparable to U.S. private voluntary organizations. The Toyota Foundation is beginning to follow the path of the Ford and Rockefeller Foundations; some informal groups contributing to community development can be found. But their role is still negligible. Business, on the other hand, plays the dominant role. Government, particularly JICA, makes a major contribution, although the separation of function between government and business is nowhere as sharp as in the U.S. case.

Japanese Business Firms

Information on the activities of Japanese business firms related to human resource development was obtained from interviews in Jakarta and its environs,
and from a detailed questionnaire that was returned by 64 firms. In addition to
our own questionnaire, which focused largely on managerial and professional
training, a survey conducted by Japan External Trade Organization (JETRO) in
1983 provided useful information on 95 Japanese firms. Unfortunately its
information on training did not distinguish between training of production
workers and of professional, technical, and managerial workers. A survey of
foreign firms conducted by AID in 1985 also provided information on the
contributions of specific Japanese firms (Cox 1985).

Questionnaire data collected from business firms has to be interpreted with
care. First, the self-selected sample for which information was obtained cannot
be claimed as fairly representative of the universe from which it was drawn. The
second reason for caution is a bias in reporting for those firms that did respond.
The number of expatriates is believed to be understated. The evidence for this is
that a number of "raids" by the Ministry of Manpower in 1984 on firms
employing expatriates uncovered hundred of expatriates without work permits.
The responses on training are believed to overstate the hours of training and
particularly the formal aspects of training programs, because of pressures by the
government of Indonesia (GOI) on foreign firms to conduct training, and also
because of the greater ease of demonstrating the implementation of formal
training, with manuals, and schedules, than of informal, on-the-job training.
Firms using more formal, structured training procedures no doubt are overrep
resented among the firms answering specific questions on training.

Human Resource Development Activities

Recruitment.   Because Japanese firms have a strong preference for promoting
from within, we asked a question designed to identify the source of managerial
and professional trainees: Does the firm hire new managerial and professional
employees directly from the outside? Twenty-four answered yes, and 35 answered no. When asked whether the firm hired directly from the outside for
positions above the entry level, a surprising 21 firms answered yes, and six no;
more than half did not answer this question. There may have been problems of
interpretation, or reluctance to admit what might be construed as "raiding",
which is frowned upon. New employees were hired locally; only one was hired
overseas.

Training.   The great majority of firms (40 of 54 reporting) train all new
employees (excluding general orientation to the firm). Of these, 35 conducted
training in their own local plant, 30 overseas, and 21 in local institutions (most

firms checked more than one answer). Trainers were local employees in 32 firms, employees of affiliated firms in 12 firms, and outsiders in 12 cases. Ten firms had space dedicated exclusively for training, whereas 34 used normal working space. Although most training was conducted during normal working hours on the job, 15 firms provided time during normal working hours devoted exclusively to training, and 11 conducted training outside normal working hours. Slightly over half the training time—for firms providing a breakdown—was on-the-job training, but a surprising 31 percent on the average was lectures and 16 percent demonstration. Only 17 firms reported hours and weeks devoted to training. Of them, nine reported two weeks or less, five between two and five weeks, and three between 10 and 25 weeks.

When asked whether the firm conducted training for its managerial and professional employees after completion of initial training and accumulation of work experience, 15 firms replied in the affirmative. Six firms indicated that their training program was open to outsiders, 31 that it was not. (Fifteen of 95 firms replying to the JETRO survey also answered affirmatively.)

In addition to training for their own employees, many firms conduct training for others; 18 reported for customers, 19 for suppliers, and 12 for others, in most cases students. (The JETRO survey did not ask whether firms provided training for suppliers or customers, but it did ask whether they provided technical advice to related companies, or assistance to parts suppliers and subcontractors; 59 firms of 95 answered yes to each question.) As to contributions to human resource development other than training, 10 provided instructors for local training and educational institutions; 14 provided equipment, supplies or facilities; 12 gave financial support to institutions; and eight offered financial support for students. Fifteen firms indicated other contributions, of which 13 turned out to be acceptance of student trainees, in most cases for very short periods. It appears that this is regarded more as public relations than meaningful training by most firms, as few chose to include these activities under training contributions to outsiders other than customers and suppliers. The remaining "other" contributions include two instances of language training and two programs for sending students to Japan. (The JETRO survey found that 52 firms were making or had made donations to schools or educational institutions, and six offered public scholarships.)

The best-known training contribution by Japanese firms in Indonesia is that of Toyota. Toyota-Astra, the distributor of Toyota products, initiated training in 1974 in motor vehicle maintenance and repair for Toyota dealers, but later made training available to others, and expanded two-week training programs to one month to a total of 500 trainees a year. In 1976 the Toyota-Astra Foundation was established, which expanded Toyota's contributions to include scholarships, gifts of laboratory facilities and teaching materials to schools, and training for small workshop mechanics. Currently, 100-200 Toyota dealer mechanics annually take the basic course taught in eight different locations, and additional mechanics take more advanced courses. Toyota-Astra's general training for outsiders includes private repair shops, fleet owners, and others. The Toyota Foundation funds training for mechanics for privately owned repair shops: 25 are trained in every course, usually given twice a year. Toyota-Astra facilities are also used for practice work requirements by students from vocational and technical schools, and for a two-week training course sponsored by the Japanese Department of Trade. Toyota Mobilindo, Toyota's manufacturing company, offers a variety of educational and training programs similar to those used by Toyota in Japan. Some trainees (10 a year approximately) are sent to Japan. Programs available to outsiders include supervisory education for general foremen, and management education for section chiefs offered in the Ministry of Education's vocational training center.

National Gobel in consumer electronics manufacturing is another leader in human resource development. It set up the Matsushita–Gobel Institute Foundation as an educational and training center, which opened a management training center in 1983. Participants in institute training programs include staff of state-owned companies and of small and medium enterprises as well as employees of the Gobel group of companies. It also accepts one or two participants from each of Indonesia's 27 provinces.

Japanese companies have cooperated with the Ministry of Industry in a program to generate and develop potential managers/entrepreneurs of small-scale industry by offering special training to new university graduates and former government officials and former employees of public establishments.

The Association for Overseas Technical Scholarship in Japan provides training for local employees of overseas Japanese firms as required by the firms sponsoring trainees. Programs range from two to 13 weeks. This and other training centers in Japan for overseas employees receive substantial subsidies from the Japanese government.

Turnover. Turnover of managerial and professional personnel is reported by 57 firms, with annual rates over the last three years showing a very wide variation. Many firms have zero turnover; 24 have less than 2 percent; 19 firms have between 2 percent and 10 percent; 14 firms report rates in excess of 10 percent; and four in excess of 20 percent. High-turnover firms are spread among various industries, although construction is represented as expected, but so is textiles; there is some concentration among metal products and motor vehicle parts. As to destination of turnover, two firms report that half is to the domestic sector, half to other foreign firms; 10 respond that all turnover is to the foreign sector and another eight that 70 percent or more is to the foreign sector; eight report 100 percent to the domestic sector and a ninth, 70 percent domestic. More than half the firms provide no breakdown. As to utilization of skills and experience gained in the Japanese firm, five firms felt that all those leaving made good use; 12 that half or more made good use of their skills; but five felt that none of those leaving made good use of their training and experience.

In spite of considerable turnover, the potential for technology transfer appears limited, as the greater part of turnover is to other foreign firms. Also, in the opinion of the responding firms, a large proportion of departing managerial and professional employees did not make good use of their training and experience in their new employment.

Skill Shortages. Overall turnover rates may conceal large variation between specialties. Firms were asked to identify problem specialties in terms of difficulty in hiring and high turnover. Eleven firms reported 19 specialties in which they had difficulty in hiring. Those that appear more than once include engineers, accountants, personnel managers, and electrical specialists (which may overlap with engineers). Specialties in which firms reported high turnover (17 firms) include engineering, accounting, and electric specialties, but also maintenance specialists. Other specialties (among which welding is listed twice) appear to have highly skilled workers not usually considered to be professional or managerial. It is not clear whether their inclusion reflects an upgrading in Indonesia given the scarcity of highly trained workers, or the Japanese failure to make the sharp distinctions by occupational group that are accepted in the United States.

Information was also obtained on skills other than professional and managerial critical to the operation of the firm and that the firm felt were not supplied by local educational and training institutions. Thirty-three firms responded, although a number of responses, such as "all professional and technical fields," were not helpful. Of the specific skills mentioned, some were highly specific to a narrow industry, and a few were really professional and managerial skills. Most often mentioned were specialists in maintenance of machinery and second, in quality control.

Fourteen firms reported that shortage of skills and adequate training facilities caused them to modify their products or production processes. One stated that it could not use the same machinery as it used in Japan, another that the level of capacity utilization was much lower. Several referred to quality control problems and suggested that they may have accepted lower quality than they would have accepted in Japan. Others did not specify product or process adjustments but simply restated their dissatisfaction with labor skills. Earlier Tsurumi (1980, 313–15) reported that some Japanese firms installed machinery and automated procedures to replace labor for lack of training and supervisory skills. Also, as firms were forced to provide their own maintenance and repair facilities for lack of adequate outside suppliers of such services, they preferred older machinery with whose maintenance they were thoroughly familiar to the latest models.

Employment of Expatriates. All but one firm employed regular expatriates (as distinguished from short-term expatriate employees). One-fourth had more than 10 expatriates, another fourth between six and 10, about one-third employed between three and five. Eleven firms out of 64 had less than three expatriates: eight had two, and two had one. This is only part of the picture, however. Twenty-six firms employed short-term expatriates, and half of those employed as many or more temporary expatriates as regular expatriates. Japanese expatriates are found on the factory floor as foremen and trainers as well as in management and top technical positions.

When asked specifically whether or not limitations on employment of expatriates affected their training effort, 19 firms replied that they did; of these, five claimed it increased their training effort, five claimed that it reduced their training effort, and nine gave such studiously ambiguous answers that it was not possible to determine whether the impact they asserted increased or decreased training activities. Foreign firms, and perhaps Japanese firms in particular, are very sensitive at this time both to restrictions on the use of expatriates and to pressures from the government to expand training activities for outsiders and to accelerate training of Indonesian workers to replace expatriates. The 1984 discoveries by the Ministry of Manpower of many expatriates working without proper work permits demonstrate that many foreign firms, Asian firms in particular, find limits on the use of expatriates excessive, although until the recent stricter enforcement they were not much constrained by them.

Domestic Content. Because most Japanese investment, particularly in manufacturing, has been for the domestic market, and would not have been made in the absence of protection from import competition, it is clear that tariffs and other restrictions on imports have contributed to human resource development and technology transfer by Japanese firms. The requirements for increased domestic content for firms already manufacturing in Indonesia have also increased human resource development and technology transfer by Japanese firms, which are the ones most affected by domestic content legislation. The impact has been less than anticipated, because Japanese firms have tended to increase their own range of manufacturing activities or persuaded other Japanese firms to produce parts and components for them in Indonesia, rather than developing or seeking out Indonesian firms. Also, as the cost of parts produced in Indonesia is in many cases much higher than the price of imports, it is easy to arrive at a high percentage value added using Indonesian prices while still importing a major part of the cost of a car using Japanese prices. In addition to the motor vehicle industry, which has been most affected by domestic content requirements (in 1986, 85 percent, and aiming at 100 percent by the end of the decade), the heavy equipment industry is also affected.

The question to which we have no answer is the possible deterrent effect of uneconomic domestic content requirements on new investment by firms not already committed to the Indonesian market, and in particular on investments with export prospects. Gray has demonstrated that protection of the motor vehicle industry, among others, generates negative value added at border prices, that is, protection consumes more foreign exchange than it saves (Gray 1982, 43–48). The technology transferred is inappropriate at this stage of Indonesia's development, although some of the human resources developed (managerial in particular) are not technology-specific.

Direct Investment.   Information on Japanese investments in Indonesia is available from BKPM, which reports approvals rather than actual investments. It excludes investments in oil, gas, and coal, and in banking and insurance, which are far less important among Japanese than among U.S. investments. Total approved investments from 1967 through 1982 were $4,344 million or 37 percent of total foreign investment approvals. (Overall, realized investments were not quite half of approved investments, although the ratio varies widely by industry and by country.) The bulk of Japanese investment approvals was in manufacturing, within which textiles was by far the largest industry, followed by metal products, motor vehicles and parts, chemicals, and wood and pulp. Although investment has been concentrated in labor-intensive industries, production is predominantly for domestic markets, and the principal motive has been preservation of markets formerly supplied via exports that were being closed by protective measures. A second motive has been development of natural resources needed for the Japanese market, natural gas in particular (Wie 1984a; Yoshihara 1978, 64–74).

In recent years, new investment applications have declined, although not as sharply as U.S. applications, from an annual rate of around 22 in the early 1970s to 7 annually for the last half of the 1970s and early 1980s (U.S. Embassy 1983, 15). There are reports of Japanese firms closing down their operations in Indonesia. Most of the reasons preferred for the decline in U.S. investments would apply to Japanese investments, although home-country legislation is not among them. Japanese firms are much more tolerant of Indonesian requirements for joint ventures and minority ownership and do not complain out loud about the costs and uncertainties of government regulation.

Licensing.   Highly aggregated data on payments for technology are available for 1980. In all manufacturing (which should account for nearly all payments), a total of $29.7 million was paid to Japan, an increase of 58 percent over 1979. Of this total, 16.6 percent was in transport equipment, 16.4 percent in chemicals, 16.2 percent in electrical machinery, 14 percent in nonelectrical machinery, and 12.9 percent in textiles. Some payments were made in construction (Pacific Economic Cooperation Conference 1983, 10, 11, 18). Nearly all payments are believed to be from affiliated firms. Given widespread practices of transfer pricing and restrictions on license fees, it is difficult to draw any conclusions about the amount of technology transfer from payments data.

Given the close relation between most of the payers and the payees, technology payments and direct investment indicate much the same thing. The ratio of technology payments in 1980 to cumulative Japanese investments in manufacturing in Indonesia is very low, well below the ratio for any other ASEAN nation. But again, because of differences in payment regulations between countries, it is not possible to infer with confidence that there is much less technology transfer per dollar of investment in Indonesia than in other countries. The inference is consistent with the fact that Japanese investment is concentrated in industries such as textiles, which report relatively small technology payments and little technology transfer.

Government Organizations

Japanese government loans and grants to Indonesia totalled $1,227 million between 1968 and 1975, of which project cooperation accounted for $711.4 million (or 58 percent) of the total; foreign exchange credits (for commodity imports), $349 million or 28 percent; and cooperation on food production, $172 million or 14 percent, of which $51 million consisted of grants. Subsequent loans, in particular oil and LNG development loans of Y118 billion and the Asashan project loan of Y61.5 billion, raised total Japanese assistance to Indonesia in loans and grants to Y971.6 or approximately $3,021 million. In 1981 total Japanese government loans to Indonesia amounted to $247 million, in addition to which the Japanese government contributed $37 million in technical cooperation and $15 million in project cooperation.

These totals tell us little about the extent of Japanese assistance because they include outright grants, loans on commercial terms and loans on varying concessionary terms. Furthermore, they include equipment, investment financing, and other contributions besides human resource development contributions.

JICA. The principal government agency involved in the transfer of technology via human resource development is JICA. It operates some training centers in Japan for trainees from LDCs, but it is primarily a contractor for services provided by other government agencies, educational, training and research institutions, and business organizations and firms (Japan International Cooperation Agency 1985).

JICA finances training in Japan both for groups (in courses reflecting the greatest common needs of LDCs), and for individuals (in accordance with specific requirements, particularly for local counterparts of Japanese experts in Japan-assisted projects). Trainees are mainly administrative officers, researchers, and middle-level technicians both from public bodies and from the private sector. Training is provided at eight centers operated by JICA and also in cooperation with various government agencies, universities, and enterprises both public and private. The cumulative total number of trainees from 1954 through 1984 was 5,379 from Indonesia, including 627 undergoing training in FY 1984. By sector, the largest number of trainees is in public works and utilities, followed by agriculture, forestry, and fisheries. Of less importance are mining and industry; planning and administration; and public health and medicine. Of least importance are commerce and tourism, human resources, and energy.

JICA also sends Japanese experts primarily to government agencies, research institutes, academic institutions, training institutes, or construction projects for planning, survey and research, guidance, and instruction. The total sent from 1958 through 1984 was 1,993, slightly more than half of whom were part of the independent experts dispatch program (similar to IESC). The total for 1984 was 351. The same sectors as in the case of trainees–agriculture, forestry, fisheries, and public works and utilities–received the largest number. Other fields receiving a number of experts were manufacturing, public health and medicine, human resources, and planning and administration. These numbers include experts assigned to international organizations in Indonesia.

Its "Project-Type Technical Cooperation" program includes both the acceptance of trainees and the dispatch of experts, and also supply of equipment and materials in an integrated manner. The types of cooperation involved include industrial development, technical cooperation center, agriculture, forestry and fisheries, as well as health, medical, population, and family planning.

JICA's annual report does not provide a breakdown between supply of equipment and materials and human resource development activities. Nor does it do so between training for professional, managerial, and technical personnel critical for technology absorption and transfer, and training for skilled and semiskilled production workers, craftsmen, and operatives. Fortunately, JICA in Indonesia responded to a survey of international donors by AID on training activities as of 1982, reporting on activities being conducted, recently completed, and about to be initiated (AID/Indonesia 1983). The major projects include a middle-level agricultural technicians training project, training 450 individuals a year for five years; a nursing education program to upgrade nursing education; and a chemical industry training and development center to train engineers and technicians, to include three-year courses for a maximum of 150 students plus short-term courses. There are projects for training seamen and loggers. Several other projects in agriculture specify only the value of equipment and do not indicate the nature of training or number of trainees. These projects are legume production, sericulture, the Lampung agricultural development plan and extension center, mariculture development, animal disease investigation centers, afforestation, plant protection, and remote sensing projects. Other projects—in biomedical research, control of communicable diseases, family planning, development or irrigation and drainage facilities, and in development of building materials—also involve some high-level training and technology transfer, although no indication is provided of its nature or extent.

JICA, in cooperation with the Department of Manpower, recently opened the Center for Vocational and Extension Service Training in Bekasi for training and upgrading instructors from the department's network of vocational training centers. This new center is part of JICA's ASEAN Human Resources Development Project, with centers in other ASEAN nations, all of which will eventually accept trainees from member countries.

JICA also conducts a development survey program that involves sending teams of experts for a variety of purposes. The program assists in regional development planning; in specific public projects; in resource surveys, feasibility studies, and project design. These surveys invariably involve valuable experience for Indonesian staff members that is impossible to quantify. The cumulative total number of survey team members sent to Indonesia from 1962 through 1984 was 5,803; the total for 1984 was 832. The largest numbers were in agriculture, social infrastructure, and transportation, followed by energy, development planning, and telecommunications.

JICA's priorities for technical cooperation with Indonesia are as follows.

1. Agricultural development, including community development and production increases, with particular emphasis on rice.
2. Energy development: both hydro and thermal power generation.
3. Transport and communications structures and facilities—roads, railroads, and ports.
4. Manpower development.
5. Medical and public health development, including family planning.

Other activities of JICA are less directly related to technology transfer via human resource development. One is equipment supply, which as a separate program was limited to the University of Indonesia, the Bureau for Technical Cooperation, and Indonesian Republic TV in FY 1984. Equipment supply, however, is part, often a major part, of the projects above mentioned, and often complementary or essential for their human resource development activities. JICA offers long-term low-interest loans to Japanese firms for improvement and expansion of facilities and for experimental projects to promote development of agricultural, forestry, mining, and manufacturing activities, as well as for social development. These loans are intended for projects which, because of their high risk or because the benefits are not appropriated by the firm, do not pay on a commercial basis.

JETRO. JETRO, initially a trade-promotion organization, has expanded its role since 1980 to include industrial development and associated training and human resource development. In cooperation with the Indonesia–Japan Enterprises Association it assists in the development of small- and medium-size firms, holding seminars on management and appropriate technologies and sending an

Indonesian mission to study Japan's subcontract and contract-out system. It promotes non-oil-and-gas exports by conducting market surveys in Japan and other countries, providing technical assistance to improve product quality, organizing buyers' and Indonesian export promotion missions, supplying commodity inspection experts, and holding product development seminars. It has also organized seminars on energy conservation and energy substitution technologies.

In sum, the Japanese government contribution to technology transfer via human resource development is closely related to the needs and activities of Japanese business. It is largely project-related and technology-specific. Training is primarily for production workers. There is a noticeable shift in orientation, however. JETRO has expanded its agenda from trade promotion to assisting industrial development and marketing. JICA's activities are less project-oriented than in the past, and make a greater contribution to increasing absorptive capacity through assistance to educational and research institutions. It appears to be placing less emphasis on the policy of cooperation, that is, mutual benefits, and more on one-way flow of aid from donor to recipient without calculating economic return to the former, viewing benefits in a wider perspective.

References

AID/Indonesia. 1983. Office of Education and Human Resources, *Joint Indonesian External Assistance Agency Training Activities as of 1 March 1982*. Jakarta.

AID/Indonesia VHP: 7/20/83. Private and Voluntary Organizations (PVO) Development Programs.

Cox, Grant. 1985. *A Survey of Private Sector Training in Indonesia*. Jakarta: AID (November).

Donges, Juergen B., Bernd Stecher, and Frank Wolter. 1980. Industrialization in Indonesia. In Gustav F. Papanek, ed. *The Indonesian Economy*. New York: Praeger.

Gray, Clive S. 1982. Survey of Recent Developments. *Bulletin of Indonesian Economic Studies* (November): 1–51.

Japan International Cooperation Agency. *Annual Report 1985*. Tokyo.

McCawley, Peter. 1979. *Industrialization in Indonesia: Developments and Prospects*. Occasional Paper no. 13. Canberra: Development Studies Centre, The Australian National University.

Pacific Economic Cooperation Conference. 1983. *Issues for Pacific Economic Cooperation: A Report by the Task Forces*. Report by the Task Force on Technology Transfer. Jakarta: Centre for Strategic and International Studies (October).

Survey of Current Business. 1985. August.

Technical Assistance Information Clearing House. 1982. *Development Assistance Programs of U.S. Non-Profit Organizations: Indonesia*. New York: American Council of Voluntary Agencies for Foreign Service, Inc. (March).

Tsurumi, Yoshi. 1980. Japanese Investments in Indonesia: Ownership, Technology Transfer, and Political Conflict. In Gustav F. Papanek, ed. *The Indonesian Economy.* New York: Praeger.

United Nations. 1982. *Transnational Corporations and Their Impact on Economic Development in Asia and the Pacific.* Economic and Social Commission for Asia and the Pacific Joint CTC/ESCAP Unit on Transnational Corporations, Bangkok.

U.S. Embassy, Jakarta. 1983. *Investing in Indonesia.* (October).

U.S. Embassy, Jakarta. 1984. *Labor Trends in Indonesia.* (May).

Wie, Thee Kian. 1984a. Japanese and American Direct Investment in Indonesian Manufacturing Compared. *Ekonomi Dan Keuangan Indonesia.* 32(1): 89–105 (March).

———. 1984b. Japanese Direct Investment in Indonesian Manufacturing. *Bulletin of Indonesian Economic Studies* (August): 90–106.

Yoshihara, Kunio. 1978. *Japanese Investment in Southeast Asia.* Monographs of the Center for Southeast Asian Studies, Kyoto University. Honolulu: The University of Hawaii Press, 1978.

5

U.S. and Japanese Contributions to Technology Transfer via Human Resource Development: Thailand

U.S. Contributions

In reporting our findings in Thailand we shall follow the same approach as for Indonesia. We will stress the differences as much as possible and avoid unnecessary repetition. U.S. organizations are again classified into three groups: private nonprofits, business firms, and government agencies.

Private Nonprofit Organizations

The TAICH country report for Thailand of April 1981 (Technical Assistance Information Clearing House 1981) lists 75 U.S. nonprofit organizations conducting development assistance programs in Thailand. Of the 75, 53 were able to provide data on expenditures in 1980, totaling $32.6 million. However, differences in fiscal year, and methods of financial reporting and of estimating dollar values of commodities, equipment, and material shipped make these totals rough approximations at best. The totals are understated because numerous organizations did not report or are not among the 75 listed. Also, it appears that not all donations in kind are included and that the value of volunteer services is not estimated. The totals can also be greatly overstated because much of the expenditures are for relief, rural and community development, and health. These expenditures, although they include education and training components, are only indirectly if at all related to technology transfer and associated human resource development as we conceive it.

Most of the organizations, including all the major contributors to human resources development, are the same as the ones we found in Indonesia.

The Rockefeller Foundation is among those foundations whose contribution is primarily toward increasing absorptive capacity. The foundation promotes agricultural development through support of teaching and research at Kasetsart and Thammasat Universities. In health and population it supports research at Mahidol University; in economics and related fields, it supports training and research at Thammasat University. It also provides support for the Ministry of Public Health's research on contraceptive methods.

The Ford Foundation supports research and training in environmental management at Mahidol University; advanced training and research in economics at Thammasat University and the Institute of Southeast Asian Studies; in water management by the Irrigation Department; and supports a population and development program at Chulalongkorn University.

The Agricultural Development Council, which is interested in management of renewable resources as well as in food production and water management, maintains a specialist at Kasetsart University and supports advanced overseas training.

The Asia Foundation's activities are quite diversified. They can be classifed more as diffusion than transfer of technology; the foundation does support some research and university training related to agricultural and rural community development, including health and nutrition. Its efforts have been concentrated in some of the more remote regions of Thailand and institutions in those regions. It is expanding its activities in management through faculty and graduate student development. It also administers AID-funded international education and training programs.

When Thailand is compared with Indonesia, we see that the nonprofits in Thailand contribute to increasing absorptive capacity by concentrating more on research. They also rely more on local educational institutions for advanced training, and less on overseas training, and use their grants as a means of allocating existing resources to priority areas as well as for expanding local resources.

IESC, whose first project in Thailand was in 1965, provided assistance to 249 projects by the end of 1982. The projects covered a wide range of activities, with machinery, metal and electrical products, transportation equipment, banking and finance, communications, transportation, and utilities accounting for nearly half the clients. Clients included several government agencies and public enterprises. A major part of IESC's recent projects are with former clients. It is diversifying its services by undertaking quick feasibility studies and reaching out to smaller firms by bringing in volunteer experts to assist two or more small firms in the same industry or with similar needs.

VITA has been active in Thailand, primarily in the area of appropriate technology for rural areas and small enterprises. It is here that VITA designed a water-pumping windmill made entirely of indigenous materials for farming use; the windmill may have a commercial potential. VITA has trained Thais in information services for technology and skills in Washington-based programs.

The FMME, which has initiated an industry- and country-specific training program in the management of technology, is cooperating with the Thailand Management Association, the Association of Thai Industries, and the Asian Institute of Technology. The U.S.–ASEAN Center for Technology Exchange, which FMME helped found, will assist ASEAN business in information on technology trends in the United States and in locating U.S. sources of training, technical assistance, technology, and investment.

Whereas IESC and VITA "retail" technology and/or firm-specific training and technical assistance, FMME "wholesales" to industries rather than to firms.

The International Institute of Rural Reconstruction, in addition to sending Thai technical personnel for training at its center in the Philippines, operates a training center in Chainat province. It fosters rural community development, promotes crop diversification and improvement of cultural practices, and public health projects.

A large number of nonprofits are concerned with improving living conditions in rural areas. The nonprofits provide training in agriculture, health, nutrition, family planning, handicrafts, and related institutional development. They build, operate, help staff, and support a large number of schools (mostly trade and agricultural), and demonstration projects. They provide teacher training; establish health facilities and train health personnel at all levels; and train local leaders for village development. Much of this effort is via volunteer services. Much of the technology introduced and training complementary to it are simple and widely available, although some of the training is for medical specialists (such as by the World Rehabilitation Fund) and for agronomists and the like. The nonprofits' main contribution is not to technology transfer so much as to adoption and diffusion of technology. They provide leadership, stimulate awareness of problems and possibilities, modify attitudes, supplement resources, and build appropriate institutions. Indirectly they contribute to technology transfer by inducing greater acceptance of new ways, and by speeding diffusion.

There is a major difference between the organizations in Thailand whose agenda is primarily rural agricultural community development, health improvement and village employment creation, and similar (many of them the same) organizations in Indonesia. In Thailand a high proportion of the organizations' attention and resources is concentrated on the refugees from Cambodia, Laos, and Vietnam who are concentrated along Thailand's eastern borders. Such aid is classified more as relief than as development. Catholic Relief Services, the largest of these organizations, with expenditures of $7.7 million in 1980 and a foreign staff of 91 (85 of whom are volunteers) and local staff of 102, is almost exclusively concerned with refugees and their needs.

The sources of funds available for nonprofits are diverse and by no means exclusively from the United States. Some organizations have their own resources, such as the Ford and Rockefeller Foundations; many rely on contributions from sponsoring organizations and individuals, such as Catholic Relief Services or the YMCA. AID contributes to a number of them, including the IESC and VITA.

Some nonprofits receive funds from international organizations, from aid agencies of other nations, foundations, and from the government of Thailand. Nonprofits, or private voluntary organizations more accurately, do not include private universities or research organizations. For example, Stanford Research Institute, whose contribution to human resource development and technology transfer may be very important, is performed under contract with AID or other public agencies and not as its institutional objective. The nonprofits considered are conducting their activities autonomously, as their primary function, with or without AID or other support.

U.S. Business Firms

Information on U.S. firms was obtained from a variety of sources. Twelve firms were interviewed. Some of the individuals interviewed have many years of experience in Thailand and could shed light on an entire industry and beyond. Their reports were supplemented by the views of knowledgeable individuals in education and training organizations, AID, and the American Chamber of Commerce in Thailand. On the basis of interviews, a questionnaire was revised and sent to some 120 U.S. firms. As in Indonesia, there are some problems in identifying U.S. firms. The Chamber of Commerce membership list includes many firms that have no formal ties to U.S. firms, and it is the only listing available. Unlike in Indonesia, wholly owned subsidiaries are permitted. Only firms with a formal corporate link with U.S. firms were included. Questionnaires were returned by 31 separate firms, although not all of them answered all the questions. This number is just over a quarter of the firms to which questionnaires were sent. It is not a scientific sample. The very diversity of the industrial distribution of U.S. firms in Thailand precludes the possibility of a truly representative sample. The findings based on the sample must be regarded only as illustrative of the behavior of U.S. firms in Thailand. We also relied on two extensive studies of foreign firms in Thailand by Mingsarn Santikarn (Santikarn 1981) and Ian McGovern of the University of Manchester (McGovern 1983).

Human Resource Development Activities.   Nearly all the firms reporting did all of their recruiting for managerial and professional openings in Thailand, and the two exceptions did most of their recruiting there. Some firms indicated some preference for Thais with degrees from U.S. universities, but the reason given for this was that study in the United States was a convenient screen for English language competence, not that training at Thai institutions was inadequate. On the contrary, the attitude toward Thai universities was generally positive. Educational requirements or expectations for new hires are sometimes higher than for similar (or even the same) firm in Indonesia.

Whereas recruitment for managerial and professional employees in Indonesia was almost exclusively at the entry level, there is considerable lateral hiring of trained and experienced workers in Thailand. Thus six of 29 firms answering stated that they do not have a training program for new hires. Those with regional headquarters in Asia conduct a good deal of their training in the headquarters city (Singapore, Manila, Tokyo, or Hong Kong), which would be the same for trainees from all countries represented. Among the 18 firms giving a time breakdown, 7 included on-the-job training, and 14 devoted at least half their training time to on-the-job training. The training program was open to outsiders in four cases. Of the 14 firms reporting the total number of hours of training for new hires, four reported two weeks or less; two reported a year or more. In nearly all cases where training is conducted for new hires (21 out of 24 firms), some training is done by local employees in the local plant. Eighteen firms conducted some training overseas (three of them exclusively overseas). Thirteen firms use outside instructors, and the same number (not all the same firms) send new hires for training in local institutions.

Continued training and updating for managerial and professional personnel is done by most firms, but the frequency and intensity depends on the rate of change of products or technologies used by the firm. Fifteen firms report further training for their managerial and professional personnel. In a number of cases training does not appear to be regularly scheduled, being instead on a need basis, or limited to particular specialties. Such training is done more systematically by firms whose entry training is regionally organized, with instructors being sent from other units of the parent organization.

Locally organized training in Thailand (whether entry-level or updating and upgrading of middle-level employees) depends on local educational and training institutions and outside instructors much more than is the case in Indonesia. This is one reason why less training appears to be done by U.S. firms in Thailand than in Indonesia.

Training of customers is common where the product is complex, and an integral part of marketing; it is also done by firms that are only distributors rather than manufacturers of the product. However, these latter firms in turn receive training and technical assistance from the manufacturing firms, often firms with no investment in Thailand. Distributors receive training in maintenance and repair as well as in operations via periodic visits of manufacturers' representatives, though rarely in the U.S. plant. Twenty of 29 respondents provide training or technical assistance for customers, and eight do so for suppliers. There is no report of training suppliers in response to domestic content pressures. Training of suppliers is an important activity of food processing firms.

A study of technology contracts (United Nations 1984, 226) found that 26 percent of the contracts with U.S. multinationals contained training provisions. Others in fact provide training and technical assistance although not specified in the technology contract examined. However, most of these contracts are with affiliated firms.

Many firms—including seven of those responding—have established ongoing programs for training outsiders. The programs are primarily for students, but in some cases local small businessmen and employees of government agencies participate. Nearly all firms make contributions to local educational and training institutions in the form of gifts, scholarships, or provision of instructors. Of the 31 firms returning questionnaires, 18 report making contributions of some kind to human resource development; seven provide instructors (others are ready to do so) to local educational and training institutions; 10 make contributions of equipment, supplies, or facilities; nine make financial contributions to educational or training institutions; and nine make contributions to individuals for study in domestic institutions (one of which also supports students in foreign educational institutions). These training activities and contributions to local educational institutions appear more extensive, more established, and somewhat more institutionalized, than in Indonesia. U.S. firms and the local institutions on which the firms rely appear much better integrated and interactive. Scholarships from firms and business organizations for education in the United States are less commonly found than in Indonesia.

All of these observations are consistent with reports on the state of higher education in Thailand. There are reported to be over 100,000 Thai graduates of U.S. universities; nearly 11,000 (compared to 8,000 for Indonesia) have received AID participant training, mainly in the United States. There is an emerging large surplus of college graduates. The quality of many of the programs relevant to business firms has greatly improved.

Turnover and Training.   Turnover of trained and experienced managerial and technical personnel from foreign firms to domestic firms is a major process of technology transfer. It also induces firms to train more workers to replace them. Turnover assumes lateral entry. High turnover rates from foreign to domestic firms further require an expanding demand. Turnover rates will also vary with specialization; some skills are industry-specific, as in petroleum exploration; others (managerial skills and finance skills in particular) are much less firm- and industry-specific. Hence one would expect higher turnover rates in such specializations.

Twenty-two of 29 responses reported hiring above the entry level; half of this group reported that half or more of their new hires were above the entry level. The greater frequency with which lateral entry of experienced workers is reported, and the absence of training programs for managerial and professional new hires in some cases, suggest a turnover rate higher than that found in Indonesia. Ten of 28 firms experienced annual rates in excess of 10 percent. As expected, the finance industry experiences higher turnover than most, but not nearly as high as in Indonesia. One does not find turnover rates as high as those in the banking industry in Indonesia, nor are there reports of large-scale "raids" (which would reflect acute needs and acute shortages). There were fewer complaints about competitors offering much higher salaries in order to attract key people, as happened in Indonesia with both domestic and foreign firms. In both countries turnover is almost invariably said to be "low" (these statements generally refer to turnover of all workers), even though there is evidence to the contrary. A rate of 10 percent is mentioned by Hongladarom (Hongladarom 1982, 93). Firms appear to be reluctant to acknowledge what might be regarded as normal turnover.

Of the 19 firms reporting on the destination of turnover among managerial and professional employees, only one reported that their workers went predominantly to foreign firms; two reported that about half went to foreign firms; most of the others reported that all turnover was to the domestic sector.

It appears that turnover in Thailand is somewhat higher than in Indonesia; that differences in turnover rates across industries are smaller; and that the more adequate supply of skills and experience allows firms to recruit at middle levels as well as at entry levels, to trade off training versus lateral entry. Thus turnover may not compel a firm to maintain and/or expand its training activities as it does in Indonesia. More of the turnover is to domestic organizations than in Indonesia. Thai firms looking for professional managers hire them away from U.S. and European MNEs. New graduates see work experience with an MNE as increasing their value (Kanchanadul 1984, 11–12).

The most common cause of turnover of managerial personnel, reported by nearly all respondents, is of Thai-Chinese called to join their family firm. The large Thai-Chinese business community, part of which is evolving from the traditional family firm and is hiring professional managers, is generating increasing turnover. Members of such family firms sometimes work for multinational firms for a time for the explicit purpose of obtaining the training and experience that they will later bring to their family's firm. On the whole the respondents thought that the departing workers (mainly managerial) would put their training and experience to good use in the family firm.

Because of the hierarchical nature of Thai society, which carries over its structure into the labor force, turnover of managerial and professional employees often involves a whole "team" associated with a senior employee, rather than just individual employees. Loyalties are transferable between employers, but remain intact within a group. Such group turnover can be a very effective method of transferring technology and associated know-how to a domestic firm, thus greatly increasing its capabilities.

A word about women. They are prominent in Thai business in managerial and professional positions. The reason is not a high status for women, but a value system in the society that gives greater prestige to government service and religious vocations (which are preempted by men). This is one reason why there is a relatively abundant pool of recruits for managerial and professional positions: Positions are open to women, and there are university-trained women available.

The high concentration of demand for technical, professional, and managerial workers in the greater Bangkok area and vicinity is a factor that promotes a well-functioning labor market for higher-level skills. Information about job openings and job candidates is more readily available locally than nationally. Job changes can be accomplished without the need for migration. There is a negative aspect to this of course, namely the lack of adequate opportunities in the more remote areas of the country with large populations. There is also the difficulty of attracting or holding such skills; this clearly impedes a wider geographic distribution of economic growth and employment opportunities.

For all these reasons, turnover of experienced managerial and professional employees to domestic organizations contributes to technology transfer, but may not induce much additional training by the firms experiencing turnover.

Skill Shortages.   Two questions shedding light on the availability of professional and managerial skills were included in the questionnaire. Firms were asked to identify skills particularly difficult to hire, and skills with exceptionally high turnover. Twelve firms indicated difficulty in hiring some specialties (excluding replies that were too general to be useful). Six firms mentioned engineers or specific engineering subspecialties; one firm cited engineer-managers; three mentioned marketing specialists; four mentioned managers; and three cited accounting and related specialties (several mentioned more than one specialty in which they had difficulty hiring). This of course does not mean that no other specialties are a problem; some are not available locally and are staffed with expatriates (oil and gas exploration, for instance). Ten firms reported on specialties with high turnover. Six mentioned engineers; four, marketing/sales personnel; one, credit officers; and one, cost accountants.

Firms were also asked whether there were skills other than professional and managerial that were critical to the operation of the firm that were not supplied by Thai educational and training institutions. Eighteen firms reported that there were none; five reported that there were, listing electronics technicians, computer operators, tire builders, and workers with paper manufacturing know-how. The list of course is limited by the industry coverage of the firms that responded. It is assumed that firms not answering this question had no problems. A larger number of firms responded on recruitment practices for scarce skills. These firms were equally divided between hiring trainees specifically and selecting trainees from among their own employees; most firms did both. Of the nine firms reporting on training procedures for scarce and critical skills, four provided overseas training, as well as training in their own plants.

When asked whether skill shortages influenced the choice of technology used or of products manufactured, four firms indicated that shortages had affected their choice of products. Pharmaceuticals in particular are influenced in their choice of products by the high training costs required. (In fact, certain production processes are considered too costly and the products requiring them are therefore not produced.)

A shortage not reflected in questionnaire replies but that was mentioned by a number of respondents is that of experienced senior managers. This reflects the recent development of modern business and industry in Thailand, and the predominance of family firms that only now is giving way to professional management. Also mentioned in the context of exports is a shortage of senior experienced marketing specialists.

Employment of Expatriates.   The number of expatriates in U.S. firms in Thailand is clearly smaller than in Indonesia. In some cases it was possible to compare the number in the same company in both countries. Senior managerial and professional positions are more likely to be filled by Thais than by Indonesians. To an unknown extent this is attributable to a longer period of continuous activity in Thailand than in Indonesia for many firms. Some companies were able to report on the number of expatriates early in the company's activities, revealing that there has been a large decline in their number. This is in conflict with McGovern's finding that there is no appreciable decline as a function of age of firm. Because the decline is mainly in the early years, it may not be revealed in McGovern's data. The most frequently observed number of expatriates in U.S. firms is one or two—suggesting that their only function is managerial control for the parent firm, and that there may be no further decline. Of 31 firms returning questionnaires, eight had no expatriates (most of these were small firms with less than 20 employees), eight had one expatriate (half had over 150 employees), five had two expatriates, and six had three to five expatriates.

Not a single respondent complained about Thai government policies on the employment of expatriates. Every firm asked was employing fewer than allowed; no firm's training activities had been affected by government policies on expatriates. Expatriate employment does not appear to be an issue with U.S. firms.

A substantial number of firms with U.S. principals, partners, senior managers, and professionals operate in Thailand, but are purely Thai; they are neither related to U.S. firms via ownership or in their decision making. Some may be distributors of U.S. products; others are not. U.S. citizens who came to Thailand in World War II, during the early days of AID, or during the Vietnam war, stayed, formed their own firms, or teamed up with Thais in founding firms. Such firms operate largely in the trade and service areas, or in consulting and technical assistance. This form of technology transfer is of limited importance, and is individual, not institutional. It is outside the realm of policy recommendations; these must be addressed to governments and private institutions. No such phnomenon was observed among the Japanese, nor for that matter among U.S. citizens in Indonesia (although there are a few individual examples). This phenomenon can be characterized as brain drain from the United States.

Domestic Content.   No U.S. firm reported any problems with requirements or pressures to replace imports with domestically produced inputs. The only reports along this line referred to motor vehicles, which now must have 50 percent domestic content. This policy is under heavy criticism because, given the small annual sales of cars, the costs are high. Although the current situation cannot be reversed, it does not appear that motor vehicle firms will be required to move further toward domestic production. Differential tariffs of course provide some encouragement to import substitution, but Thailand appears relatively permissive in this regard.

Pharmaceutical firms however report themselves faced with what they regard as unfair competition from domestic competitors, in particular from the Government Pharmaceutical Organization. This body seeks to monopolize government hospitals and other markets for pharmaceuticals that are subject to public administrative influence.

U.S. Direct Investment.   The main agents for technology transfer at Thailand's stage of development are foreign enterprises, which introduce new products and processes, generate demands for new and higher skills, and contribute significantly toward supplying these skills by training and support for Thai educational institutions.

The U.S. Embassy conducted a survey of U.S. investment in late 1983. (Thai-American Business 1984, 29–31). The estimated value of U.S. investments in Thailand in 1983 was $3,094 million, of which nearly half, $1,538 million, was in oil and gas, $613 million in manufacturing, $832 million in banking and finance, and $111 million in other activities. (The *Survey of Current Business* [1984] estimates, which reflect capital outflow from the United States only, were $729 million for the same year, of which $564 million were in oil and gas.) The average U.S. ownership share was 77.5 percent. However, only $913 million constituted funds brought in from outside. Major manufacturing investments were in chemicals and pharmaceuticals, consumer products, metal products, food and beverages, paper, and rubber processing and tires. The electronics industry is expanding rapidly. Although there is no breakdown of total sales between domestic and foreign markets, only 30 percent of the firms produced exclusively for the domestic market; 5 percent produced exclusively for export; and the remaining 65 percent for both. (Twenty-six of the firms returning our questionnaire provided market breakdowns: three produced exclusively for export, eight had some exports, and 15 produced exclusively for the domestic market.)

Thailand has only one-third Indonesia's population, and no known vast reserves of oil. It does not offer Indonesia's future market potential. But it has a more adequate infrastructure for business, a better supply of well-trained workers at all levels, particularly at the technical and managerial level, and a less intrusive government. U.S. investment is perceived in Thailand as far below desired levels (Mason 1978, 25). Thailand appears to have unexploited potentials for competitive export industries. A number of firms interviewed are intended to serve a regional market, but this market (Vietnam, Cambodia, Burma, and in one case, Bangladesh) is at present either closed or of minimal value.

The question as to why there is not more U.S. investment, or more that is export-oriented, was not pursued. However, the complaints of various U.S. businessmen may shed some light on this issue. The most common complaint was about U.S. law, specifically the Foreign Corrupt Practices Act, for the reports and testimonials required of Thai partners and contractors, and the reporting requirements from U.S. firms. But this is only a recent development, and it is not believed to be a major deterrent, certainly no more so in Thailand than elsewhere. There was a widespread feeling that government administration was weak and inconsistent, with rules sometimes changed retroactively, hence unpredictable, but otherwise not unreasonable. The system of "grease" payments was regarded as effective and cheap. These complaints do not appear to reflect major deterrents to U.S. investments, only annoyances.

Most U.S. investment is not "promoted" by the BOI. Because so little U.S. investment is taking advantage of the tax and other concessions offered by BOI, it appears possible that there is a problem in providing information to potential investors. Feasibility studies conducted by the Thai government or contracted out by it do not appear to offer adequate information for potential investors. The questions of how particular products are selected for feasibility studies, how adequate the studies are themselves, and how the process of communicating with potential investors functions merit some investigation.

Government competition does not appear to be a factor in deterring investment, except in pharmaceuticals. Government has been reducing its industrial activities, and has sold some public firms to the private sector. There is legal protection for patents since 1979, and for trademarks, but violations occur and implementation of the laws may be too slow and inadequate for effective protection. The pharmaceutical industry in particular complains that it faces unfair competitive conditions in drugs manufactured by the Government Pharmaceutical Organization in violation of patents and trademarks (this industry is excluded from patent protection). Thailand is arriving at the point where credible protection for industrial property rights is important for investment in some industries and for some products that otherwise find Thailand to be an attractive location.

Licensing and Fees.   The annual survey conducted by the U.S. Department of Commerce of fees and royalties paid by foreign affiliates to U.S. firms indicates that total payments rose rapidly from $9 million in 1977 to $30 million in 1983. Subtracting oil, which was not listed in 1977 and grew rapidly in 1982 and 1983, the increase was from $9 million to $18 million. Looking at royalties and license fees only (excluding service charges and rentals for tangible property and film and TV tape rentals), the total rose steadily from $4 million in 1977 to $9 million in 1983. More than half of this was in manufacturing, and within manufacturing, chemicals accounted for more than half, followed by the food industry and other manufacturing. In the last two years (1982–83) trade has been almost as large as manufacturing. (The 1977 benchmark survey reports total payments of $8 million, $3 million of which is for royalties and license fees.) These totals are not complete, but are believed to represent the right order of magnitude. As indicators of technology transferred, they leave much to be desired.

Data from the RTG give total remittances of management fees, patent royalties, and technical assistance fees to the United States. These amounts have been rising steadily from slightly over $3 million in the 1973–75 period to $12.9 million in 1980, and $26.5 million in 1981 (no reason is given for this extraordinary one-year increase) (United Nations 1984, 195, 199).

There are too many differences among these diverse sources in their sampling, concept, or coverage to allow for reconciliation. Because in any case they are very poor measures of technology transferred or complementary training, what matters in the above estimates is that the amounts are rising rapidly, but are still low.

Mingsarn Santikarn made a detailed study in 1975 of 388 technology contracts involving 257 foreign firms. Of 49 technology contracts with U.S. firms examined, 14 were in chemicals (mainly paints and cosmetics), six each in electrical products and textiles, and five each in construction materials and metal products and equipment. Most contracts were associated with imperfect markets and goodwill; even those in textiles and apparel were associated with trade names; and many involved batch mixing and assembling techniques, that is, they involved limited technology transfer (Santikarn 1981, 117–19, 130ff.). For all contracts examined, without breakdown by nation, nearly three-quarters of the remittances were intrafirm, that is, from affiliates or joint ventures. The wide range of fees paid, and the large number of restrictive conditions found in many contracts, render the totals paid of limited value as indicators of the amount of technology transferred, or of training required (United Nations 1984, 205 ff., 226, 230).

The evidence, such as it is, suggests that most technology transfer is to affiliated firms, and is associated with investment in Thailand by foreign firms; transfer to independent domestic firms is not yet a major factor.

Government Organizations

Most of the U.S. government contribution to technology transfer is routed through AID, which, however, is mainly a funding agency and clearinghouse for information, decision making, and contracting; it is a performer only to a limited extent. The actual technology transfer may be through other government agencies, land-grant colleges, other educational institutions, and through contracting firms that train workers to construct projects and to operate them and maintain them once built.

In the 1950s and 1960s the main contribution of AID was toward building a physical infrastructure (roads, ports, airports, communications, power, and irrigation). This involved large-scale training of Thais in the skills necessary to build the projects, and to operate and maintain them. This effort included Thai companies (which acquired the experience and resources to become contractors), and government agencies (which had operational and maintenance responsibilities). AID also invested on a smaller scale in the educational infrastructure—vocational education, medical facilities and education, public administration, and teacher training. (Over 11,000 Thais have been sent abroad for advanced training, mainly in the United States.) In combination, these efforts greatly increased Thailand's capacity to absorb new technologies, although the educational aid was not, except for vocational training, technology-specific.

Current AID projects are focused on agricultural-rural development, mainly in the Northeast. Most of the projects are no longer large physical infrastructure investments, but small-scale demonstration projects in water management, development of new agricultural products, rural health and community development. The training and technical assistance component does involve substantial technology transfer to government agencies involved in rural-agricultural development of the Northeast and to some agribusinesses (seed companies), but some of the assistance is best characterized as an extension service. Leaders are taught to reach and persuade rural residents to adopt widely known and relatively simple technologies, and to demonstrate in key areas what can be done, with the expectation of widespread imitation. Total funding in Thailand is running slightly over $30 million a year.[1] Employment of U.S. citizens in Thailand declined from 580 in 1960 to 22 in 1983. Part of this decline however constitutes technology transfer. Thai nationals assumed the responsibilities formerly discharged by U.S. citizens.

In April 1984 a separate Office of Agriculture was established within AID. Its portfolio of ongoing projects at that time totaled $60.3 million, most of it in projects with a five-year duration. They included research and development at Khon Kaen University, agricultural development planning, several model demonstration projects of integrated rural development, land settlement, watershed development, small-scale irrigation, rain-fed agricultural development, and a seed development project. An agricultural technology transfer project for $5 million was in the planning stage. Expenditures are primarily for technical assistance, research, and training.

The Office of Agriculture will be working closely with the Bureau for Science and Technology of AID, which has a number of research and development projects active in Thailand. These projects include improved seed industry development; improved varieties, production, and utilization of soybeans and peanuts; improved postharvest systems for grains; storage and processing of fruits and vegetables; water management; soil management support; agriculture; coastal resources management (proposed); weed control; dryland management; and nitrogen fixation. As these are centrally funded projects, conducted in numerous countries, no expenditure data are reported for Thailand. They involve technical assistance from land-grant universities, and collaborative research and training in Thailand.

An agreement on cooperation in science and technology between Thailand and the United States was signed in April 1984. The agreement is intended to encourage and facilitate direct contracts and cooperation between government agencies, universities, research centers, and private sector institutions and firms of the two countries. Cooperation will be undertaken in agriculture, health, energy, space, environment, natural resources, and any technology relevant to agricultural and industrial development. One underlying objective is to build upon the base of advanced scientific training (much of it in the United States) by strengthening institutions and programs utilizing and advancing the knowledge of Thai scientists and engineers. Priority areas are projected to include biotechnology, metal technology, resource use and management, electronics, and preservation. AID will develop a major science and technology project and has formed a council of Thai leaders to advise it (Briskey 1984, 24–26).

AID also provides one-quarter of the core budget of a number of international research institutes with representation in Thailand, of which the best known is the International Rice Research Institute (IRRI), other international centers include the Asian Vegetable Research and Development Center, the International Crops Research Institute for Semi-Arid Tropics, and centers for corn, wheat, tropical agriculture, aquatic resource management (AID 1984).

Much attention has been given to the collaborative Japanese–United States–Royal Thai Government project in Northeast Thailand as the single example of Japanese–U.S. cooperation in aid to an LDC. The AID–RTG agreement calls for $2 million in grants from AID for research development at Khon Kaen University over a six-year period starting in December 1983. The objective is to improve the quantity, quality, and relevance of research activities and services relating to problems facing Northeast rural communities by strengthening the capability of Khon Kaen University's Research Development Institute. Most of AID's contribution is a research fund for farming systems and rural development research, with lesser amounts for U.S. technical assistance, local consultants, training, and publications.

The Japanese-sponsored Northeast Agricultural Development Project on the other hand chose cooperation with the Department of Land Development rather than Khon Kaen University. The project consists mainly of constructing and equipping a research facility. Thus there are not one but two mutually supporting, parallel projects.

AID's recent greater emphasis on the private sector has led to direct support (through funding and technical assistance) of business and business-serving organizations. The Office of Private Enterprise is providing technical assistance to the BOI for feasibility studies and promotional efforts. The office also finances staffing for the Thai Chamber of Commerce, Association of Thai Industries, and the Thai Joint Agricultural Consultative Committee. The office supports the Institute for Management Education in Thailand Foundation, the IESC, and provides financial and technical assistance to seed associations. It has made loans to local banks for relending on a long-term basis to small and medium agrobusinesses in rural areas, to farmer cooperatives and for land development and irrigation projects.

Mention should be made of the Asian Institute of Technology (AIT) which, although an international educational and research organization, is located near Bangkok and is making a significant contribution to Thailand in technical education and research. AID played a major role from AIT's beginning in 1959, contributing $20 million over the years. Currently it provides a $3.125 million scholarship fund for ASEAN nationals.

In the 1960s and early 1970s, during the war in Vietnam, the Department of Defense made a contribution toward infrastructure development with civilian uses. DOD built ports, roads, and airfields, established communications, and trained thousands of engineers, managers, and technicians. The current contribution is on a much more limited scale. U.S. assistance to the Thai military takes two forms: International Military Education and Training (IMAT) and Foreign Military Sales (FMS). The sales of military equipment includes training as part of the package. The amounts vary widely depending on the volume of sales and their composition, from $400,000 to nearly $6 million in recent years. The amount was $2.6 million in 1983. Training is mainly in communications electronics and aircraft maintenance, and some flying operations. It is estimated that 60–70 percent of FMS training has a civilian counterpart. The main components of IMAT are resource management, operations management, and technical training, nearly all of it conducted in military schools and colleges in the United States. The United States provides tuition and school support costs, whereas Thailand provides transportation and living allowances. The number of officers trained is expected to rise from 300 in fiscal year 1984 to 350 in 1985.

In addition to AID, various government agencies contribute to a limited extent through training programs with diverse funding sources, and with technical reports. The Department of Commerce and the Department of Agriculture both have training programs for LDC nationals, including Thais. Other departments have discontinued similar activities. The Fulbright Commission has sent nearly 800 Thais for graduate study in the United States and some 300 U.S. academics for teaching and research in Thailand. The activities of OPIC have been reported on page 91.

Contributions of Different Agents Compared

U.S. firms have contributed to firm-, industry-, and technology-specific training and in particular to management experience. They also contribute to the demand for relevant education, without which there would have been less improvement in absorptive capacity.

As in Indonesia, AID's contribution to technology transfer via human resource development has been quite different from that of business. Its technology-specific training contribution has been in building and operating the nation's transportation, communications, water management, health, electric power systems, which are primarily the functions of government. Apart from this, it has contributed to institution building in education, to the training of faculty for teaching and research, to providing a labor force with a greater capacity to absorb new technologies, and to some extent to modify and adapt technologies developed elsewhere. The major foundations have followed a policy of educational institution building in areas related to development potential, including administration, agricultural sciences, medical sciences, engineering, and economics.

However, both physical infrastructure and institution building in education are largely completed, and both AID and some nonprofits have scaled down their activities in these areas. AID and most of the nonprofits involved in training with some technology transfer element remain concerned with the rural-agricultural problems of the Northeast. These problems now involve institution building and behavioral changes at the village level, with the leverage of the physical infrastructure and the educational and training institutions already in place. With agricultural productivity still low, and without opportunities to extend the cultivated area, there is obvious need and opportunity for business involvement in the commercialization of farming, in the Northeast in particular. Agribusiness has a role both in providing the market for new products, and in providing the inputs for higher productivity.

Japanese Contributions

Japanese Business Firms

Information on the human resource development activities of Japanese firms was obtained from a number of sources. In addition to interviews conducted in field trips to Bangkok and environs, a detailed questionnaire stressing professional and managerial training was sent and 84 questionnaires returned. Useful information was also obtained from a survey conducted by the Japan Overseas Enterprises Association in 1983; from the ongoing work of McGovern based on intensive interviews with a very large number of foreign firms; and from the work of Santikarn focusing on technology agreements.

Human Resource Development Activities

Recruiting.   Because Japanese firms, unlike U.S. firms, maintain no sharp distinction between managerial and professional employees and others, we asked a question not addressed to U.S. firms: Does the firm hire managerial and professional employees directly from outside? Thirty-nine firms answered yes, 40 answered no. This is a higher proportion than found in Indonesia; so is the proportion of firms that hire for positions other than entry-level: 35 reported that they did, only five that they did not. (Thus our first question does not tell us what proportion of entry-level trainees are hired from outside rather than selected from among current employees.) The source of managerial and technical employees, and the extent to which firms may hire experienced workers, would affect the extent and nature of their training activities. Hiring was done locally by all but one firm, which hired a single employee overseas.

Training.   The question whether all new managerial and professional employees received training, other than general orientation to the firm, elicited a surprising answer: Almost equal numbers (40 and 38) replied affirmatively and negatively, respectively. Absence of training in many firms reflects in part the hiring of experienced workers above entry level. It may also reflect the satisfaction that many Japanese firms report (JOEA) with the training provided by local institutions.

Of 39 firms supplying information on the locale of training, 36 provided training in their plants, 26 overseas, and 18 in local institutions. All firms providing training in their plants used local employees as trainers; 13 also used employees of affiliated firms; only two used outsiders as trainers. Training procedures were generally informal: only three firms had space dedicated exclusively for training; only seven scheduled time exclusively for training. The 32 firms reporting on the allocation of training time among various procedures on the average devoted 20 percent to lectures, 59 percent to on-the-job training, and 15 percent to demonstrations. (Most firms did not, most of them probably could not, provide such breakdowns and it is assumed that their training was predominantly informal and on-the-job). As to length of training, which was requested in terms of hours per week and number of weeks, most firms did not answer and additional firms replied that they did not know, which is what one would expect from a largely informal, on-the-job process. Of the 19 firms giving specific answers, 10 reported a training period of a week or less, four of two to four weeks, and five of five to 12 weeks. In comparison with Indonesian replies, training of managerial and professional workers in Thailand is more informal, of shorter duration, and relies less on outside trainers.

Only two of the firms conducting training reported that their programs were open to outsiders. On the other hand, 28 firms provided training for suppliers, 19 for customers, and 14 for others, primarily students.

A total of 38 firms reported making or having some contribution to human resource development other than training. Nine provided instructors for local educational or training institutions; 14 provided equipment, supplies, or facilities; 10 made financial contributions to local educational and training institutions; and seven provided financial support to students. Nineteen firms reported other contributions; 15 accepted student trainees; two provided services for sending students to Japan; and two offered language training. Acceptance of student trainees apparently is not regarded as training in the same light as that provided for employees and others. Perhaps this training was considered information and public relations, and possibly as useful for screening for potential employees. The evidence for this is that of these 15 firms, only seven reported conducting training for outsiders other than suppliers and customers.

The Japan Overseas Enterprises Association survey of Japanese firms provided some additional information on trainees, trainers, and training content. It revealed that Japanese firms frequently used Thai organizations for training employees, including the Technological Promotion Association (Thai-Japan); the Personnel Management Association of Thailand; the Thailand Management Association; the Institute for Skill Development (Ministry of Labor); and the Employers' Confederation of Thailand. Most of the companies indicated they were satisfied with the training. All respondents, however, sent some trainees to Japan, particularly managerial and supervisory staff.

The survey classified workers into rank-and-file, foremen, and middle managers, and asked the firms whether or not they conducted training for them in a number of technical and business and managerial subjects. Most did not train rank-and-file workers. The subjects with the highest proportion of positive responses were production technologies and quality control, followed by safety control and "welfare." A higher proportion trained foremen, with more than a third training them in all four technical areas: production, productivity, safety, and quality control. Nearly a third trained foremen in accounting, finance, sales, purchase, and international trade. The percentages were much higher for middle managers: over 40 percent for each of the technical subjects, but over 50 percent for each of the business and managerial subjects, and 73 percent for objective control and managerial control.

Turnover. Information on the annual turnover rates of managerial and professional employees between 1980 and 1983 was reported by 71 firms. Thirty-three reported turnover rates of less than 2 percent; six reported rates of 2–4.9

percent; six reported rates of 5–9.9 percent; 17 experienced 10–19.9 percent turnover; and nine firms experienced over 20 percent turnover. Looking at the industrial distribution of firms with high turnover rates, most textile firms report high rates, although in manufacturing, paint, and food products firms report the highest rates. Very high rates are reported by some firms in finance, trading, and services; and moderately high rates are reported by construction and transportation firms.

Much of the turnover was to other foreign firms. Of 34 firms providing a breakdown by destination, 14 reported that all their turnover was to foreign firms, four that 80–90 percent was to foreign firms, and nine that turnover was equally divided between the foreign and domestic sectors. Only six firms reported that all their turnover was to the domestic sector, and a seventh that 80 percent was domestic. When asked what proportion of departing managerial and professional employees made good use of their training and experience in their subsequent job, opinion was divided. Six firms felt that less than half of departing employees made good use, seven that half did, and 11 that 80–100 percent of departing employees made good use of their experience.

Skill Shortages. Firms were asked to identify managerial and professional/technical specialties in which they experienced difficulty in hiring. Those listed by more than one firm included engineers (several subspecialties were mentioned), technical designers, personnel managers, financial and accounting specialists, and sales and marketing personnel. We also asked firms to list specialties in which they experienced higher-than-normal turnover. Mentioned most frequently were engineers/technicians (the Japanese do not draw as sharp occupational distinction as the U.S. firms, and it is not always clear what level of technical training or capability is intended); second, accounting; and third, managers/foremen (again, there is a boundary problem in responses). There were multiple responses listing "professional" and "technical" without further specification.

We also asked firms to identify critical nonmanagerial, nonprofessional skills that were not being supplied by the Thai educational and training system. A fair number of answers appear to refer to professional and managerial occupations, especially engineers. There were many replies specifying three areas: quality control, production control, and production planning. Specific skills were less frequently mentioned.

It is apparent that Japanese firms had difficulty in responding to questions referring to occupational classifications. The Japanese management system places less stress on educational preparation, and more on learning by doing; it moves employees around to perform a variety of tasks instead of practicing a rigid functional classification as is the U.S. custom.

When asked whether shortages of critical skills not supplied by Thai educational and training institutions affected choice of products or technologies, 15 firms said yes, but most of the replies did not refer to specific skills. Five mentioned problems of quality control; one mentioned maintenance and servic-ing. One firm reported dropping a product, and two said they were unable to introduce products. Two firms were unable to introduce home-country technol-ogies, one of whom referred specifically to computer-controlled machinery; a third adapted its production methods to circumvent skill shortcomings.

Employment of Expatriates. All 84 firms returning questionnaires employed regular expatriates. Eleven firms, however, employed only one; 20 employed only two; 32 employed between three and five; 12 between six and 10; and nine firms employed more than 10 regular expatriates. The total was 390, or an average of 4.64 per firm. In addition, however, 34 firms reported employing temporary short-term expatriates totaling 246, or an average of 7.2 per firm, although the distribution was highly skewed: 19 firms employed one or two temporary expatriates; six employed between three and five; four had between six and 10; and five employed more than 10.

Examining the distribution of expatriates by industry, high numbers are found in some construction/consulting firms and trading firms as one might expect. Somewhat surprising is the fact that six of 10 manufacturing firms with seven or more regular expatriates are textile firms. They happen to have large numbers of employees, but otherwise it is not apparent why they should have so many expatriates. Looking at the use of temporary expatriates, on the other hand, most of them are in the motor vehicle, electrical equipment, and glass industries, which require much greater technical knowledge than textiles. It is possible that temporary expatriates are being used to circumvent restrictions on the number of regular expatriates.

When asked whether the restrictions on the use of expatriates affected their training activities, five firms replied that they reduced their training activities, two that the limitations had led them to train replacements, and two others complained about visa problems without indicating the impact on training. Only eight other firms replied to this question, and claimed no effect.

Domestic Content. Thailand uses tariffs as a means of encouraging investment and protecting domestic production. Most Japanese investment has been for the domestic market and tariffs were an important if not necessary condition for their commitment. Requirements that domestic production have a given mini-mum percentage of domestic content is another matter. This requirement raises costs of production directly, whereas tariffs only permit high-cost producers to survive (especially as in most cases imported input requirements can be brought in duty-free). Domestic content requirements can be particularly burdensome for industries that might have export prospects.

Of the firms queried, 49 replied that 50–100 percent of their sales were domestic, another five reported an even split, and 14 reported that 55–100 percent of their sales were to foreign markets. Although 10 firms reported sales to affiliated firms, only three of them sold 40 percent or more of their output to affiliated firms.

The one area in which domestic content legislation has had an impact on technology transfer and human resource development by Japanese firms is the motor vehicle industry, which is dominated by Japanese firms. The requirement that domestic content equal 50 percent of value has led to production of numerous parts, components, and accessories that would otherwise have been imported. Much of this is done by other Japanese firms, rather than by Thai contractors. The large number of temporary expatriates in Japanese firms in the motor vehicle industry may be a reflection of domestic content requirements as well as of limitations on employment of regular expatriates. The fact that the response has been largely to use Japanese suppliers rather than to develop domestic firms limits the substantial technology transfer potential of the legislation. It appears to have no deterrent effects on other investments, because extension of domestic content requirements is not being proposed.

Direct Investment.   Japanese cumulative direct investment in Thailand was an estimated $1,702 million in 1983. Most of this investment was made in the 1960s and early 1970s; by 1974 total cumulative investment was $1,544 million, close to the 1983 figure. This figure, however, greatly understates the importance of Japanese investment, for two reasons. First, most Japanese investments involve minority Japanese ownership, and second, Japanese firms are characterized by a high debt-to-equity ratio. Japanese investments were primarily in manufacturing, and constituted of large numbers of investments—the average amount being only $200,000 (current prices at the time of investment) (Yoshihara 1978, 78). This is in part because the Thai domestic market was quite limited at the time most investments were made. Within manufacturing, textiles was the dominant sector of Japanese investment in 1977, accounting for 45.5 percent of paid-up capital; chemicals accounted for 18.4 percent, and automobiles, 13.8 percent. Other industries accounting for significant investments were iron, steel, and food products. (Lee & Hongladarom 1984, 24). In terms of employment in Japanese firms, textiles accounted for over 60 percent.

A survey of Japanese investors in Thailand conducted by Somsak Tambunlertchai found that the investment incentive listed as very important by the largest number of respondents was tariff protection. It was followed by exemption (or reduction) of import duties on imported materials. Other factors listed by nearly half the respondents as very important were income tax exemptions, favorable conditions for remission of foreign exchange, and existence of knowledgeable local investment partners (cited in Lee & Hongladarom 1984, 26, 30).

Licensing and Technology Payments.   Technology payments consist primarily of royalties, but also include technical fees, management fees, and trademark fees. Japan received an average of 40.5 percent of the technology fees over the period 1973–80, and 33.5 percent in 1981, or $20.4 million. In 1981 technology fees were involved in 36.3 percent of the total number of technology contracts, nearly all of which were in manufacturing, with transport equipment the leading industry (presumably a consequence of domestic content requirements for motor vehicles). Other industries with 10 or more technology contracts include industrial chemicals; other chemical products; textiles; electrical machinery and appliances; machinery, except electrical; and food products. Because Japanese firms on the average charged lower fees than Western companies, their share payments may understate their relative contribution. However, it was also noted that affiliated firms paid higher fees on the average than independent licensees, and Japanese technology contracts are predominantly with affiliated firms. Forty-two percent of contracts contained training provisions (United Nations 1984, 198–99, 203–4, 212–13, 226).

Because there are no restrictions on licensing agreements in Thailand, fees may be a better indicator of the amount of technology transferred than they are in countries imposing restrictions. Nevertheless, given the numerous ways in which licensors may be compensated by affiliated licensees, and the absence of a perfectly informed market for technology, not too much weight should be placed on technology payment figures.

Government Organizations

The current Japanese aid program is running at around $350 million a year. Most of it is project-related and for equipment and supplies, although most projects involve some human resource development relevant to technology transfer. Some programs, however, have such development as their primary aim.

JICA. JICA's training program in Japan has included 5,289 Thais from 1954 through 1984, with 585 Thai trainees in 1984. The sector with the largest number of trainees was agriculture, followed by medicine and public health, public administration, manufacturing, telecommunications, human resources, and transportation. The great majority participated in group training offered to LDC nationals, with smaller numbers in individual training and (Japanese expatriate) counterpart training.

The expert dispatch program has sent a cumulative total of 2,335 experts (298 in 1984), with medicine and public health by far the largest sector, accounting for close to half the experts, followed by agriculture, human resources, and telecommunications. The total includes an unknown number sent to international organizations in Thailand, including the Asian Institute of Technology, the Economic and Social Commission for Asia and the Pacific (ESCAP), the Southeast Asian Fisheries Development Center, and the Asia–Pacific Telecommunication Association.

A total of 3,972 survey team members have been sent to Thailand since 1962, and 544 in 1984. They were concentrated in agriculture, transportation, public works, social infrastructure, and energy. Lesser numbers were active in telecommunications, manufacturing, and medicine, and public health. Although training is not their purpose, considerable incidental training of Thai technical and professional workers is involved.

Cost estimates in yen are available for 1982, with a level of activity somewhat lower than for 1984. Converting yen into dollars at the rate of 200 to 1, costs of development survey teams were $9 million, of dispatch of experts slightly less, of sending Thai trainees to Japan, under $3 million. Equipment supply cost $5 million, and other unspecified technical cooperation brings the total to $28 million (JICA 1984).

Technical cooperation center programs in Thailand were the Institute for Skill Development in the Northeast of Thailand, the King Mongkut Institute of Technology, the Furniture Industry Development Center, and a primary health care training center, all of which except the last were due to be completed by 1983. In health, JICA has completed a program with the National Cancer Institute, and promotes provincial health services. It continues a family planning program. In agriculture, it supports research and development, extension, and mechanization programs at Kasetsart University, a weed research project, a coastal aquaculture development project, research and training in reafforestation, an animal health improvement program, and an irrigated agriculture development project.

The Northeast Agricultural Development Project, in collaboration with AID and the Royal Thai Government, is the one example of U.S.–Japanese cooperation. Japan is building an Agricultural Development Research Center for the Department of Land Development, adjacent to Khon Kaen University. A grant of approximately $5 million will finance construction and equipment of the center and some research equipment and transportation vehicles for the university. AID on the other hand is working through Khon Kaen University, mainly through research funding and consultants, including technical assistance to the center.

JETRO. JETRO, in addition to its trade promotion and public relations activities, conducts surveys and collects information on resources, investment climate, markets and products that are useful for Japanese investors. It conducts surveys on management and labor conditions and provides consulting services for Japanese establishments. Its Bangkok office, which is a regional center for Burma, Laos, Cambodia, and Vietnam, surveys the conditions of the maritime industry in ASEAN countries.

Technological Promotion Association. The Technological Promotion Association was founded in 1973 with the financial support of the Japanese government, but it is operated and controlled by its Thai members. Its central objective is the advancement of technology among its members and dissemination of technology among the general public. Toward this end it conducts seminars and training sessions concerning technology, collects, edits, translates, and publishes books and documents dealing with technology. The association maintains a library of technical publications and cooperates with other institutions with similar objectives, especially those in Japan. It conducts language classes in Thai and Japanese.

Note

1. One problem with assessing the contribution of government (and international) organizations is in comparing grants and subsidized loans. AID in Thailand estimated the grant component in its loans at 71 percent, with average loan terms of 2 percent interest for 10-year grace period, then 3 percent for 30 years. (For Japan, with a five-to-seven year grace period, then 3–4 percent interest for 23 to 25 years, and 0.5 percent service charges required immediately, the grant component was estimated at 50 to 63 percent; for the World Bank it was only 7 percent).

References

AID 1984. *Program of the Office of Agriculture, Thailand.* Bangkok (May).

Briskey, Ernest J. 1984. Science and Technology: The Shape of Things to Come *Thai–American Business* (March–April): 20–27.

Hongladarom, Chira. 1982. The Structure of the Thai Economy and Its Implications for Industrial Relations in Thailand. In Chira Hongladarom, *Comparative Labor and Management: Japan and Thailand.* Bangkok: Thammasat University Press.

JICA (Japan International Cooperation Agency). 1984. *Annual Report 1983* Tokyo.

Kanchanadul, Veeravat. 1984. The Role of Multinational Firms and Its Impact on the Business Environment. Paper presented at a Conference on the Role of Multi-National Corporations in Thailand. Organized by Thammasat University, Cholburi, Thailand, 7–9 July.

Lee, Eddy, and Chira Hongladarom. 1984. Employment and Labor Relations in Foreign Companies. Paper presented at a Conference on the Role of Multi-National Corporations in Thailand. Organized by Thammasat University, Cholburi, Thailand, 7–9, July.

McGovern, Ian. 1983. Managerial Aspects of Technology Transfer in Thailand. Paper presented at Seminar of Technology Transfer, Transformation and Development, Bangkok, Chulalongkorn University, 2–6 September.

Mason, R. Hal 1978. *Technology Transfers: A Comparison of American and Japanese Practices in Developing Countries.* Pacific Basin Economic Study Center Working Paper Series no. 7, Graduate School of Management, UCLA.

Santikarn, Mingsarn. 1981. *Technology Transfer: A Case Study.* Singapore University Press.

Survey of Current Business. 1984 (August).

Technical Assistance Information Clearing House. *Development Assistance Programs of U.S. Non-Profit Organizations: Thailand.* New York American Council of Voluntary Agencies for Foreign Service, Inc. (April).

Thai–American Business. 1984. U.S. Investment in Thailand Survey (March–April): 29–31.

United Nations 1984.*Costs and Conditions of Technology Transfer Through Transitional Corporations.* ESCAP/UNCTC Joint Unit on Transnational Corporations, Economic and Social Commission for Asia and the Pacific, Bangkok, (April).

Yoshihara, Kunio. 1978. *Japanese Investment in Southeast Asia.* Monographs of the Center for Southeast Asian Studies, Kyoto University. Honolulu: The University of Hawaii Press.

6

The U.S. and Japanese Contributions Compared

There are significant differences between the Japanese and U.S. contributions to the objective of human resource development for technology transfer. Some of the differences are related to the different approaches between the two governments to the issues of technical cooperation. Some differences are caused by the particular attitudes and policies of business firms from the two countries in regard to the strategy of operating overseas, including the policy on human resource development. Those differences can be interpreted also as a reflection of different circumstances, experiences, and concepts of modernization and industrialization.

After 1950, Japan also experienced very large structural changes that required training and retraining of much of its labor force. The more modest pace of structural change in the United States could be accommodated in greater part by new entries into the labor force, in particular given the enormous increase in college enrollments during the 1950s and 1960s. The fact that formal educational programs in management, business administration, and the like are well-established in the United States, but not in Japan, also helps explain differences in approach.

Agents

In terms of agents, a significant contribution is made by U.S. private nonprofit organizations whereas little is done by Japanese counterparts. Private voluntary organizations have a long history and play a larger and more varied role in the United States than in most countries.

As to government agencies, the principal agencies, JICA and AID, differ sharply in the composition of their aid. JICA's contribution to human resource development is nearly all associated with specific projects, principally infrastructure projects (transportation, power, and utilities), and agricultural and natural resource development and processing projects. AID contributes primarily to increasing absorptive capacity rather than to industry-specific skills through high-level training largely overseas for government officials, university faculty, administrators in agricultural and health sectors, or to overseas education in development-priority areas in addition to project-associated training. Its contribution in terms of industry projects and industry-specific skills is limited mainly to small-scale projects that are part of agricultural-rural community development programs. It is difficult to compare the magnitude of JICA's and AID's contributions because figures are not disaggregated adequately between human resource development and other contributions. Comparison is difficult because no line can be drawn between training activities that constitute technology transfer, and training activities that constitute an addition to the contribution of domestic schools and training by production organizations. The lack of comparable information is indicative of different views as to the method and content of manpower training for the purpose of technology transfer.

Behavior of U.S. and Japanese firms in Indonesia and in Thailand by and large conformed to the differences generally observed and noted in chapter 1, but with some exceptions. Contributions by business firms differ in amount and in composition. In terms of investment, U.S. contributions are somewhat larger larger than Japanese if one includes oil and gas, much smaller if this sector is excluded. In terms of composition (excluding oil and gas), U.S. investments are relatively large in services, including banking, finance, and trade, and, in Indonesia, consulting and contracting services. Japanese investments are more concentrated in manufacturing. Within manufacturing, Japanese investments are concentrated in textiles, metal products, and motor vehicles and parts in both countries; the United States emphasizes chemicals, rubber, electrical and electronic equipment, cement, and pharmaceuticals.

Characteristics of Business Investment

Japanese firms are more willing to accept minority ownership than are U.S. firms. Because of their greater reliance on expatriates, Japanese firms are in a better position to maintain managerial and technical control with minority ownership than U.S. firms. The use of expatriates, however, appears independent of the Japanese ownership share. To the extent that an LDC joint venture is dependent on imports from affiliated firms for equipment and components, or if it supplies its output to affiliated firms in Japan or the United States for further manufacturing and/or marketing, the ownership share becomes unimportant for control, and even unimportant for the earnings of the parent firm. The concentration of Japanese business in LDCs in old industries and technologies partially accounts for their acceptance of minority positions.

Japanese firms appear more closely linked as suppliers to joint venture subsidiaries in Thailand and Indonesia than U.S. firms and, except for electronics, are more important as sole customers for their products. There may also be a difference in investment strategy, with Japanese firms more interested in market share and U.S. firms in profits and rate of return, which also leads Japanese firms to greater acceptance of ownership sharing and minority ownership.

Japanese firms keep equity investment low, and rely more on loan capital than firms of other nationalities (cheap credits from Japan are a factor). In many cases the equity investment of the Indonesian joint venture partners is financed by loans from the Japanese partner, which are tied to purchases of machinery and other inputs at above-competitive prices from designated suppliers in Japan (Wie 1984, 101–2).

It was not possible to compare the import content of Japanese and U.S. firms nor to determine the extent to which high import content is used for noncompetitive transfer pricing practices. The differences in industry structure, as well as limited information, precluded a direct comparison. There is some information on the import content of Japanese firms. In Indonesia in 1974, 46 percent of inputs for Japanese firms in Indonesia (44 percent for textile firms) came from parent firms in Japan (Wie 1984, 104). In Thailand, the import dependence of Japanese firms is reported as 60 percent for materials and over 90 percent for capital equipment (Tambunlertchai & McGovern 1984, 39). In neither Indonesia nor Thailand are Japanese firms export-oriented. They are reported to export 16 percent of their output in Indonesia and 20 percent in Thailand, much of it textiles and apparel in both countries. In this regard the Kojima-Ozawa thesis (Ozawa 1979) is not supported; there is little difference between Japanese and U.S. firms, which also are predominantly supplying the domestic market. The industrial composition of Japanese investment, however, has a large representation of the older, more labor-intensive industries, although recent investment is shifting the composition of Japanese business more toward the U.S. structure.

Training by Business and Its Contributions to Technology Transfer

The main agents of technology transfer are business firms, and the main processes are the training of Indonesians and Thais in the skills needed for the transfer of technology; the accumulation of experience by trainees; and their turnover to domestic organizations that can put their acquired skills and experience to use. There are many differences between the typical behavior of Japanese and U.S. firms and in the consequences for the host country.

Before describing them, let us note that the U.S. business community in Indonesia and in Thailand is dominated by large multinational firms, whereas the Japanese business community includes many smaller firms whose entry and operation is facilitated by trading companies or *sogoshosha*. This difference in composition between the U.S. and the Japanese business communities accounts for some of the observed difference in behavior.

First some warning about the limitations of the data for comparing Japanese and U.S. firms. The industrial distribution of Japanese and U.S. firms is so different that matched Japanese and U.S. pairs of firms (matched by industry or type of product) are not representative of either Japanese or U.S. companies. In fact only a few matched pairs could be obtained from questionnaire data. Nor can it be claimed that the firms returning questionnaires are an unbiased sample of either U.S. or Japanese firms. The qualitative comparisons between Japanese and U.S. firms reported below, based on interviews and questionnaire returns, are believed to be valid. However, quantitative conclusions cannot be drawn, because no sample can be equally representative of firms of both countries.

Numerical comparisons between U.S. and Japanese business refer to Thailand, where it proved feasible to distribute identical questionnaires to firms from both countries. It is also possible to compare Japanese firms in both countries. Data on U.S. firms in Indonesia are based mainly on interviews and on questionnaires distributed by AID. AID questionnaires were not limited to U.S. firms, and were not identical to those sent to Japanese firms in both countries (Cox 1985). In the case of Indonesia there are incentives to understate the number of expatriates, as evidenced by the discovery of around a thousand expatriates working without permits in early 1984. Responses on training may overstate the more formal aspects of training or even the duration of training, because of pressures by the government of Indonesia on foreign firms to conduct training, and because of the greater ease of demonstrating the implementation of formal than of informal training. These incentives are greater of Japanese than for U.S. firms, because of their greater dependence on expatriates and on informal methods of training.

There are differences in recruitment. The Japanese process tends to be much more thorough, emphasizing the social qualifications as well as the personal ability of a candidate. This process scrutinizes a candidate's ability to fit in and work as an acceptable member of a work group, on the likely loyalty to the firm, and on the probability of remaining with it. The U.S. firms place much more stress on formal educational qualifications and on performance on objective aptitude tests for the skills required. These differences have implications for future turnover. They reflect differences in the concept of work organization and manpower management.

There is little difference in hiring locale. All but one Japanese firm hired exclusively in Thailand, and all but two U.S. firms did so as well, although both did 80 percent of their hiring in Thailand. The great majority of firms from both countries also reported hiring managerial and professional employees above the entry level. (Although the question was intended to identify lateral entry of experienced individuals, it is possible that some Japanese replies interpreted the question to mean hiring of individuals with more than the minimum qualifications for the position.) With regard to Japanese firms, this is contrary to the common belief in the practice of promoting to entry-level managerial and professional positions from within. Roughly half the Japanese firms reported hiring for such positions directly from the outside, the other half from among their own employees. This may be a response to the higher-than-desired turnover rate, a response implying that not all promotions in Japanese firms are from within.

There are numerous differences in training procedure and training content, again reflecting different concepts of manpower management.

The most surprising finding about Japanese firms is that almost equal numbers (40 and 38) replied respectively that they did, or did not, provide training for new managerial and professional hires. This is consistent with hiring from outside rather than promoting from within, and hiring experienced workers above the entry level. Among U.S. firms replying, 23 did and six did not. Nearly all firms (U.S. and Japanese) conducting training did so in part in their own plants. A somewhat higher proportion of U.S. firms also conducted training overseas, and a substantially higher proportion in local educational/training institutions. The Japanese are much more likely to send new employees for training in Japan; U.S. firms rarely send any to the United States, but many send some new hires to regional centers in the Pacific. The Japanese will send production workers (and in at least one case, the entire work force) for training overseas; the U.S. firms will only send managerial and professional hires. This is a plus in terms of the quality of training, but a minus in that the training capability is overseas, not in the host country.

As to trainers, nearly all firms of both nationalities used their own local employees. Two-thirds of U.S. firms also used employees of affiliated firms as trainers, whereas only one-third of Japanese firms did so. The largest difference, however, is in the use of outside instructors. Just over half of U.S. firms used them; only two Japanese firms did so.

Training by U.S. firms is primarily for new hires. After this initial training cycle is completed, and substantial experience acquired, some managerial and professional employees who are expected to advance in the organization receive further training. It is at this stage that employees are likely to be sent to the parent firm headquarters in the United States. Firms with relatively static technology may engage in little training beyond the initial cycle except for those employees selected as candidates for advancement.

Japanese firms are said to consider training as a continuous process throughout the tenure of the employee. From the time of first entry most employees go through various steps of training according to the schedule prescribed for each category of workers. This difference is based in part on the expectation of long-term employment with the firm, so that investment in continuous training appears worthwhile. It is also based upon the policy of nearly always promoting from within, whereas U.S. firms are more likely to hire experienced workers for middle-level positions. It is also part of the group bonding social experience in the firm. Thus workers in Japanese firms are likely to receive training and experience in a variety of skills.

The replies to the question of whether managerial and professional employees received further training after accumulation of work experience in the firm appear to contradict the above observations. Only one-quarter of Japanese firms, but half of U.S. firms, answered in the affirmative. But this difference is believed to reflect in part the greater importance of multinational corporations in the U.S. responses, and to some extent the greater rate of change in products and/or processes of U.S. firms. It also fails to reveal differences in the proportion of experienced managerial and professional workers who receive additional training; the proportion is believed to be higher among Japanese firms. In fact, a number of U.S. firms indicate that further training is highly selective, and is given as needed, rather than as a standard procedure. The finding that a substantial proportion of Japanese firms do not provide follow-up training is consistent with hiring above the entry level and a higher-than-desired turnover rate.

The training process in Japanese firms is predominantly informal and on-the-job, probably more so than in Japan itself. This process involves much one-to-one interaction with instructors and experienced workers—one reason for the large number of Japanese expatriates and their presence at the shop floor level. U.S. firms are likely to conduct more formal training, with a larger proportion of the time devoted to lectures and audiovisual presentations. U.S. firms make much greater use of outside instructors, formal educational institutions, and published materials. Training content is more general, oriented to abstract knowledge, and technology-related, compared to Japanese operation-specific and situational training, much of which can be described as firm-specific acculturation. This makes training in U.S. firms more transferable to other firms and industries, in part because the training can be exhibited in the form of outlines, lectures, readings, film, and tape. The content and outcome of training are more verifiable—certifiable in the eyes of others. This is another factor in expectations of higher turnover rates than in Japanese firms.

These differences are explained by various reasons. One is the types of trainees. Japanese firms train all categories of workers, including rank-and-file operators and technicians who are not likely to have an aptitude for formal lectures and classroom work. U.S. firms are more concerned with managerial and professional personnel who are accustomed to learning through abstract media. Another reason is the content of training. A large number of Japanese firms are in manufacturing where the training tends to be machine- and production-process-related know-how; U.S. firms are in service industries where necessary knowledge is gained easily through formal training.

When training for managerial and professional workers is examined alone, sharp differences between firms were found in training procedure, as expected, but not in all regards. Only three Japanese firms had space dedicated solely for training, and all of them used regular working space, whereas nearly half of U.S. firms reporting had space dedicated solely to training and six did not use regular working space. As to training time, only two Japanese firms conducted training outside regular working hours, whereas 30 percent of U.S. firms did. The majority of U.S. firms allocated time exclusively for training, whereas only 18 percent of Japanese firms did. On the other hand, the training approach was not significantly different. The percentage of time devoted to lectures was actually lower among U.S. firms, and the percentage to on-the-job training lower, than for Japanese firms. It should be stressed that neither national sample can be regarded as representative, and that the observed differences are too small to be meaningful.

Although differences in training procedures (lectures, demonstrations, and on-the-job training) for managerial and professional hires do not appear to be significant between Japanese and U.S. firms in Thailand, contrary to common belief, a much higher proportion of U.S. than of Japanese firms provided information on hours and weeks of training. Nearly half of U.S. firms replying to the questionnaire provided such information in Thailand, whereas less than 30 percent of Japanese firms in Indonesia, and less than 20 percent in Thailand, provided such information. Among those reporting, only two Japanese firms reported more than 10 weeks of training, none as many as 25; five U.S. firms (from a much smaller total) reported more than 10 weeks, and two reported more than 25 weeks. When allowance is made for the great majority of Japanese firms that could or in fact did not report on hours of training, it may be that commonly held differences between U.S. and Japanese approaches to training professional and managerial workers are not far off for Indonesia and Thailand.

Comparing training time for managerial and professional new hires in Japanese textile firms, we find surprisingly that firms in Thailand spent considerably more time in training than in Indonesia. This is surprising because of the better preparation of workers in Thailand for such positions. This finding, however, is consistent with the much lower use of expatriates in Thailand than in Indonesia. An inference is that the Japanese firms in Thailand are training Thais for more senior and responsible positions than similar firms in Indonesia.

On training contributions other than for their own employees, there were substantial differences associated with the differences in the nature of products and services. A larger proportion of U.S. firms—20 of 27 reporting—conducted training for customers (versus 19 of 50 Japanese firms). A much smaller proportion of U.S. firms—eight versus 28 Japanese firms—reported training for suppliers, and a similar proportion (7 U.S. versus 14 Japanese firms) reported training for other nonemployees.

There are significant differences in contributions to human resource development by means other than training. A higher proportion of U.S. than of

Japanese firms made some contribution: 58 versus 45 percent. Half of the U.S. firms contributing made financial contributions to educational/training institutions, and the same number made financial contributions for students; less than a quarter of Japanese firms did. The most frequent Japanese contributions were equipment, supplies, and facilities; 14 of 38 firms reported these contributions. However, a substantially higher proportion of U.S. firms listed this category, as well as supplied instructors to local educational and training institutions. Nearly half the Japanese firms listed "other" contributions (and 14 firms checked only this category), whereas none of the U.S. firms did. Nearly all "other" replies referred to student training and briefing.

The amount of technology transferred is a function of the types of products or services produced by foreign firms, and of the production processes used. Accordingly, firms were asked whether skill shortages and the costs of overcoming these shortages by training were a factor in the choice of technology used or of products (services) produced. Twenty-two Japanese firms, all but two in manufacturing, replied in the affirmative. This was in substantial contrast to U.S. firms, only four of which replied in the affirmative (three in manufacturing, two of which were in pharmaceuticals).

On a related question, whether there were skills other than managerial and professional critical to the operation of the firm that were not supplied by the host country's educational and training institutions, there was again a clear difference between Japanese and U.S. responses. Thirty-one Japanese firms claimed that there were such skills (42 said no), whereas only six U.S. firms (five in manufacturing) answered yes (18 answered no). This difference cannot be accounted for by differences in industry mix but is believed to be attitudinal and expectational.

Turnover

The turnover rate of managerial and professional employees of U.S. firms appears to be somewhat higher than for Japanese firms. It is not clear, however, that this difference is a performance characteristic, attributable to differences in recruitment practices, in training, or in attitudes toward turnover. U.S. firms are more heavily concentrated in the high-turnover industries (high-turnover for reasons unrelated to the presence of U.S. firms). These firms provide services in general (banking and finance in particular) with skills that are less firm- and industry-specific than is the case of manufacturing industries. (In Indonesia, they are involved in the engineering contracting industry, the nature of whose business involves high turnover.)

Turnover in Japanese firms, however, is higher than we had been led to expect by early interviews. One-third of the firms reporting in Thailand, and one-fourth in Indonesia, had annual turnover of managerial and professional employees of 10 percent or more, some much more. There is also a large difference in the destination of turnover, which should be much less biased by industry mix differences than total turnover rates. Only three U.S. firms reported that half or more of their turnover was to other foreign firms, whereas 27 of 34 Japanese firms answering the question asserted that half or more of their turnover went to other foreign firms. The question arises whether many of these went to affiliated Japanese firms or not, and to what extent the turnover should not be counted, but regarded as reassignment, but no information was obtained on this point. As indicated in chapter 1, it is turnover to the domestic sector that contributes to technology transfer.

Promotion

There are important differences in promotion practices. Japanese firms consider seniority as one of the basic factors of promotion because it implies one's experience and familiarity with the company—familiarity with its work requirements, mode of operation, and organizational capacity. Stress on seniority is a factor in the relative slowness, or reluctance, to promote local employees to responsible positions. To the extent that this reluctance prevents faster development of potential among local employees, it hampers technology transfer. U.S. firms, in contrast, promote on the basis of superior performance and promise, as the achievement standard is an important principle of U.S. management. This is one factor in the greater willingness of U.S. firms to promote local workers to responsible senior positions: They have a freer hand than the Japanese in choosing whom to promote. But there is little question that U.S. firms are more willing to promote and delegate authority and responsibility to host-country employees. In fact much of their training beyond the new hires cycle is directed at preparing and testing individuals for such positions, more than imparting specific skills. (To the extent their turnover is higher, they have more opportunities for promotion.) The senior positions to which host-country citizens are advanced offer more valuable experience than in the case of Japanese firms because of the greater independence of U.S. affiliates from parent company supervision.

Perceptions about these differences between U.S. and Japanese firms may lead to some self-selection among candidates for managerial and professional positions in U.S. versus Japanese firms. They may well affect turnover rates as well, but no evidence was obtained on this point.

Both higher turnover and greater willingness to provide experience in senior positions contribute to technology transfer.

The Role of Expatriates

There is little question that Japanese firms employ more expatriates and replace them more slowly than U.S. firms. This is a pattern noted in many countries (Sim 1978, 6; Mason 1978, 16). Expatriates perform two roles: They train local workers, including their own replacements, and they perform managerial and operational functions. As trainers they contribute to technology transfer; as preempters of senior positions they reduce technology transfer. Japanese firms both send more workers for training in Japan, and send more expatriates to the host country. This implies a greater training capability than U.S. firms. Japanese firms undoubtedly supply more training at least in the form of informal on-the-job training. But they also display a greater preemption of positions at middle levels which in U.S. firms are filled by local workers as quickly as possible. U.S. replacement of expatriates is attributable to the rapid turnover of expatriates as well as to the desire to promote native workers. The short stay (two or three years) of the typical U.S. expatriate compared to that of a Japanese counterpart (four to six years) reduces training and managing contributions.

The fact that Japanese firms in LDCs are more export-oriented than U.S. firms (which invest mainly to supply the domestic market in the host country) could be a factor in the former's greater use of expatriates and reluctance to replace them. Cost and quality control are essential in production for export. But this is not significant in Indonesia or Thailand, where Japanese firms are almost as much local-market oriented as U.S. firms. Another factor is domestic content legislation, which results in production costs so high and quality problems so serious that measures analagous to those necessary for export industries may be taken. Japanese investments in both countries, and particularly in Indonesia, have been affected, particularly in motor vehicles. The cost of locally produced parts and components is so much higher than of imports that there must be strong pressures for cost reduction and quality control as a result, and consequently greater use of Japanese expatriates.

Even allowing for all the considerations above, the Japanese use more expatriates, and they are almost exclusively Japanese, whereas many expatriates working for U.S. MNEs are not U.S. citizens. There appear to be three reasons, not mutually exclusive: (1) limited knowledge of English by many higher-level Japanese expatriates; (2) ethnocentrism; and (3) the Japanese management style. This difference shows up even in companies such as IBM, whose management style is claimed to be a (or the) model emulated by the Japanese. This suggests that the difference in use of expatriates is not a matter of management style, but ethnocentrism, lack of English, or both. Because affiliates of U.S. MNEs have much more autonomy than affiliates of Japanese firms, one would expect them to have more expatriates at the managerial level. Thus the differences in the use of expatriates is even greater than indicated by the numbers available. When asked whether host-country government policies on employment of expatriates

Table 6–1
Japanese Expatriates: Thailand versus Indonesia

Industry	Country	Number of Firms	Number of Expatriates	Total Employment	Professional & Managerial Employment	Professional & Managerial Employment as % of Total Employment	Ratio of Expatriates to Total Employment	Ratio of Expatriate to Professional & Managerial Employment
Textiles & Apparel	Thailand	11	6.3	1221	62.4	5.1%	0.52%	10%
	Indonesia	8	10.8	903	23.8	2.6	1.2	45
Chemicals	Thailand	16	2.9	159	15.3	9.6	1.82	19
	Indonesia	13	7.5	385	44.5	11.6	1.94	16.8
Machinery & Transportation Equipment	Thailand	5	5.4	304	17.6	5.8	1.78	30.8
	Indonesia	8	14.8	500	19.8	4.0	2.95	74.7

Source: Questionnaires to Japanese firms collected by the authors in 1984.

affected the firm's training activities, 16 Japanese firms replied in the affirmative, in sharp contrast with U.S. firms, none of which answered in the affimative.

Adjustments to Local Conditions

The behavior of both Japanese and U.S. firms in Indonesia and Thailand conforms somewhat, with the exceptions above-noted, to the differences between them that have been observed elsewhere. But there is adaptation to local conditions. Training does not appear to be different. In fact the use of Japan as a training center by Japanese firms, and of regional centers by the U.S. firms, suggests that there is little adjustment in this area. U.S. firms—the same firm in Thailand and in Indonesia in some cases—appear to make some adjustment in recruiting. Firms that in Indonesia recruit for professional and managerial positions on overseas campuses, recruit locally in Thailand. They are also likely to have higher educational expectations: In some cases, they prefer graduate training instead of an undergraduate diploma.

U.S. firms appear to have fewer expatriates in Thailand than in Indonesia, and to have a larger representation of local employees in senior positions. One would expect this from the better educational situation in Thailand, although part of this difference is attributable to the greater average number of years in active operation in Thailand than in Indonesia.

When one compares Japanese firms in Thailand and Indonesia, clear differences appear (see table 6–1). Three groups of firms were compared, with substantial representation in both countries: textiles and apparel, chemicals, and machinery and transportation equipment. The first is more homogenous than the other two, but nevertheless all three are homogeneous enough to make comparisons worth making. (Other industries were omitted because of inadequate representation in one of the countries or obvious heterogeneity between countries.)

The differences are as follows:

1. The number of expatriates in every case is much greater in Indonesia than in Thailand.

2. Adjusting for differences in average employment by firm, the ratio of expatriates to total employment is much lower for Thailand in textiles and apparel and in machinery, and slightly lower in chemicals.

3. The ratio of expatriates to total managerial and professional employment is much lower in Thailand for textiles and apparel and machinery, and slightly higher for chemicals.

4. The ratio of total managerial and professional employment to total employment is lower in Thailand for chemicals and machinery, but much higher in textiles and apparel.

It is clear that the Japanese firms use fewer expatriates in Thailand than in Indonesia; they do adapt to local availability of managerial and professional workers. Because Thailand has a much better supply of engineering and related graduates, as well as a much better supply of managerial personnel (particularly Thai-Chinese), the differences are as expected. There is one anomaly: the high ratio of managerial and professional to total employment in the Thai textile and apparel industry. It appears that there is a greater export orientation than in Indonesia, and the composition of products and particularly their fashion-sensitivity require a higher ratio of managerial and professional employees, and greater flexibility in production, than in Indonesia.

The difference in use of expatriates between Thailand and Indonesia by Japanese firms cannot be attributed to differences between the two countries in the age of Japanese firms. McGovern has found little relation between age of firm and use of expatriates in Thailand (McGovern 1983, 6). And it is the reverse of what would be expected on the basis of policy differences between Thailand and Indonesia. Far more Japanese firms in Indonesia than in Thailand (30 versus 16) report that regulations on the use of expatriate workers have affected their training activities.

Manufacturing firms returning questionnaires (there were not enough observations to disaggregate by industry) were regressed on the year in which firms started operations. For both Thailand and Indonesia a statistically significant but very small effect of age was observed. But when textile and apparel firms of both countries were pooled, and regressed against age, the results were not statistically significant. Not much weight can be given for the results for all manufacturing, given its heterogeneous composition. Nevertheless, all the evidence, such as it is, indicates little or no reduction in the number of expatriates with age of the firm. One qualification is in order. There are very few observations for firms established for five years or less. It is possible that during this initial period there is substantial reduction in the number of expatriates.

Comparisons between Japanese responses in Thailand and in Indonesia are more appropriate than comparisons between U.S. and Japanese firms. The industry mix is quite similar for Japanese firms in both countries; a number of responses are for the same firms in the two countries. However, the sample of returned questionnaires is not a scientific one, and some differences do exist in the industry distribution of Japanese respondents in the two countries. Small differences in responses will be ignored as not significant, but large differences will be noted below.

Japanese firms are somewhat less likely to hire managerial and professional workers from outside, rather than promoting from within, in Indonesia than in Thailand, and are more likely to train all new hires. (However, firms in Thailand are somewhat more likely to provide additional training for employees with work experience in the firm.) They are substantially more likely to use outside instructors in Indonesia than in Thailand. This is believed to reflect the organization of a number of industry-wide training institutes, in which the Japanese have played a major role. Also, training programs in firms open to or aimed at nonemployees are a response to government pressures as well as to the inadequacies of most government training institutions.

In training procedures, Japanese firms in Indonesia are more likely to have dedicated training space, to schedule time exclusively for training, and to conduct training outside normal working hours in Indonesia. They also rely somewhat more on lectures. A smaller proportion of Japanese firms provide training for suppliers in Indonesia than in Thailand, no doubt reflecting the fact that they have not progressed as far in import substitution and/or that they are more likely to replace imports via diversification or reliance on other Japanese firms in Indonesia.

During the 1980–83 period, annual turnover of managerial and professional personnel from Japanese firms is very substantially higher in Thailand than in Indonesia, as indicated below.

Turnover	*Thailand–Japanese Firms*	*Indonesia–Japanese Firms*
20% and over	10	5
10–19.9%	17	9
5–9.9%	6	9
0–4.9%	32	39
Total	65	62

This difference reflects a more developed labor market in Thailand, and perhaps also a different attitude toward turnover among Thais.

Perhaps the most dramatic difference in Japanese behavior between countries is in response to the question of whether they hire managerial and professional workers above the entry level. In Thailand, 35 of 40 firms replying said yes, whereas in Indonesia, only 26 of 60 firms replying said yes. This difference reflects in part a response to higher turnover in Thailand, and a better supply of trained and experienced managerial and professional labor. To the extent this question was interpreted as meaning above minimal educational requirements, it reflects the better availability of educated Thais.

With regard to skills critical for the operation of the firm, other than managerial and professional skills, a higher proportion of Japanese firms in Indonesia than in Thailand (59 percent versus 48 percent) reported that there were such skills that were not provided by the local educational or training institutions. In training such workers, a smaller proportion of Japanese firms in Indonesia relied on domestic training institutions and a larger proportion provided training overseas. Turnover of such workers was around the average for all local employees in Thailand in nearly half the firms; below average for half (27 and 29); and above average for 10. In Indonesia, turnover was below average for 31 firms; average for 14; and above average for four. Nearly half of all Japanese firms in Indonesia responding to the questionnaire agreed that shortages of critical skills other than managerial and professional had affected their choice of production process and/or of products and services, whereas little over a quarter of firms in Thailand agreed. Nearly half in Indonesia also asserted that policies on the use of expatriates had affected their training activities, whereas less than one-fifth in Thailand agreed. The last difference may reflect domestic policy on expatriates more than skill shortages.

Other Differences

A difference in response to domestic content requirements has been reported by one source. The Japanese firms are likely to respond through other Japanese affiliates and joint ventures as suppliers, the U.S. firms more through independent suppliers that receive technical assistance as required. This is difficult to assess, however. Japanese firms are more heavily in industries—motor vehicles in particular—most affected by domestic content requirements, where replacement of imported parts and components is demanding in terms of technology and skills. On the other hand, in some areas, such as textiles and apparel, where Japanese firms are heavily concentrated, U.S. firms are absent. In some cases, U.S. firms have been displaced by large retailers, who contract directly with domestic firms, providing them with necessary assistance to meet quality requirements.

The differences between the government-business relationship among the Japanese and U.S. firms in Thailand and Indonesia can be illustrated by the difference in the experiences of the Japanese and U.S. researchers. The Japanese researcher was adopted by his embassy, which made all contacts and appointments for him and provided transportation. The U.S. researcher received helpful suggestions on likely sources of information, but was entirely on his own, although several embassy and AID officials made calls individually in his behalf. As to the questionnaire for distribution to Japanese and to U.S. firms, the Japanese questionnaire was taken over by Japanese government and business organizations in both countries, who handled the distribution, collection, and did much follow-up calling to urge return of completed questionnaires. In the U.S. case the embassy and AID, for the sake of impartiality, keep hands off as matter of policy, and business organizations are reluctant to assist independent research. Businessmen reportedly have low response rates even to queries from their own embassy. Attitudes of individuals ranged from warm support to clear opposition. Business and government appear to maintain correct if somewhat distant diplomatic relations, but lack common purpose or much influence on one another.

References

Cox, Grant. 1985. *A Survey of Private Sector Training in Indonesia*. Jakarta: AID (November).

Mason, R. Hal. 1978. *Technology Transfers: A Comparison of American and Japanese Practices in Developing Countries*. Pacific Basin Economic Study Center Working Paper Series no. 7, Graduate School of Management, UCLA.

McGovern, Ian. 1983. Managerial Aspects of Technology Transfer in Thailand. Paper presented at Seminar of Technology Transfer, Transformation and Development, Bangkok, Chulalongkorn University, 2–6 September.

Ozawa, Terutomo. 1979. *Multinationalism, Japanese Style: The Political Economy of Outward Dependency*. Princeton: Princeton University Press.

Sim, A.B. 1978. Comparative Management Practices and Performance of American, British, and Japanese MNCs in Malaysia. Paper presented to the Malaysian Economic Association Seminar on Foreign Investment: Benefits and Responsibilities, Kuala Lumpur, 10 July.

Tambunlertchai, Somsak, and Ian McGovern. 1984. An Overview of the Role of MNCs in the Economic Development of Thailand. Paper presented at a Conference on the Role of Multi-National Corporations in Thailand. Organized by Thammasat University, Cholburi, Thailand, 7–9 July.

Wie, Thee Kian. 1984. Japanese Direct Investment in Indonesian Manufacturing. *Bulletin of Indonesian Economic Studies*. (August): 90–106.

7
Conclusions and Recommendations

In earlier chapters we described briefly the environment for technology transfer in Indonesia and in Thailand, national priorities, and national policies affecting technology transfer. Then we described the activities of U.S. and Japanese organizations contributing to technology transfer and to absorptive capacity. This chapter takes a forward look: our assessment of both opportunities for and obstacles to technology transfer, and our general and specific recommendations for U.S. and Japanese organizations to increase their contributions to technology transfer to Indonesia and Thailand via human resource development. Many of our findings and recommendations are widely applicable to LDCs, not just to the two we studied.

Our recommendations represent our best judgment on those ideas suggested by others or that came to us in the process of examining the activities of U.S. and Japanese organizations and of speculating on how best to meet needs and exploit opportunities. They should be regarded as suggestions for consideration rather than as a definitive or detailed agenda.

Priorities for Technology Transfer

Technology transfer is not an end in itself, although occasionally it appears to be so viewed by some policymakers. Its general objective is economic development and growth. But this is a desideratum, not a program, plan, or policy. Other objectives, complementary with growth (and possibly correlated to it or sometimes conflicting with it), are more concrete.

Thailand sees its main problem as that of poverty and underemployment in the outlying provinces, the Northeast in particular. Its major development effort for years ahead is centered on the heavy industry complex on the Eastern Seaboard, based on newly developed natural gas fields. Because the foreign exchange savings and economic viability of the complex have been adversely affected by the steep drop in oil prices, private funds have not been forthcoming as planned, so that either public funds must be diverted from other areas, or plans must be reconsidered. Agriculture, with low yields, limited use of fertilizers and of irrigation, offers excellent prospects for growth in output and exports. However, Thailand and the United States already dominate international trade in rice, the demand for which is very inelastic; the EEC has restricted imports of cassava, so that growth in productivity must be accompanied by further diversification of crops.

Thailand faces a growing surplus of college graduates, many of them very well prepared, although many also in specialties—education, social sciences, and humanities—for which the supply of graduates greatly exceeds their present and probable future demand. A literate work force and substantial educational and training capacity offer possibilities of new export markets and rapid structural change.

Thailand has underutilized absorptive capacity for new technology, in terms of its educated manpower resources and ability to expand them. It also has a well-developed banking and financial system that could accommodate a large increase in investment (there are problems with longer-term credit, especially to small business). The country has a functioning stock market that could provide more equity capital and a secondary market for shares. Transport, communications, utilities, and business services generally are adequate and should not be a serious constraint to a large increase in investment.

Indonesia's overwhelming problem is continuing entry to the labor force of large numbers of poorly educated workers and a large and growing problem of unemployment and underemployment. It is also facing up to a deep cut in oil prices, which threatens to undermine its investment plans, and has already resulted in postponement of major projects. Forests, fisheries, and minerals offer prospects for expanding exports and diversifying away from oil and gas. An inadequate physical infrastructure and poor quality of education, which cannot be quickly corrected, limit international competitiveness in labor-intensive products for the time being and hamper employment creation for the domestic market as well. Indonesia's primary need is to improve absorptive capacity (particularly managerial and entrepreneurial capacity) for implementing conventional technology to provide employment.

But the main determinant of demand for college-educated workers is per capita income, and the level and composition of demand associated with it. Much of the employment of college graduates as incomes rise is not in export industries or in industry at all, but in private and public services, for business and industry of whatever kind, and for households. It would be a mistake to design a policy for employing Thailand's educated labor surplus in terms of encouraging specific industries that are relatively human capital-intensive, because such a policy aims only at a part, possibly a small part, of the demand for educated workers. A policy aiming at high economic growth, free to consider diverse sources of economic growth, would be not only more cost-effective, but would have a larger potential impact on employment of college-trained workers. However, with the present rate of output of universities, no attainable rate of growth can do more than slow down the growth in the number of unemployed or misemployed university graduates. To the extent that the diploma is the important product of education rather than the education itself, a growing surplus of college graduates is not a serious problem. But clearly many graduates want more from their education than a diploma, and it is questionable that a diploma alone can long retain much status value if its past correlate of access to desired positions and careers is seriously eroded.

Both Thailand and Indonesia are richly endowed in natural resources. In the case of Indonesia, oil and gas dominate value of output and export earnings but employ very few people. The major policy question is the allocation of the public revenue from the export sales of oil and gas. In the case of Thailand, exports are dominated by the agricultural and agricultural processing sector. This sector employs over 70 percent of the population; development of this sector offers the prospect of increasing employment and income directly in agriculture and indirectly in agricultural product processing and support activities.

Indonesia lags well behind in development of an economic infrastructure for business. An uncertain regulatory environment and poor information are further constraints for private investment. But the shortage of managerial and entrepreneurial skills is the principal bottleneck in absorptive capacity.

Thailand is concerned with regional imbalances, specifically development of the Northeast, decentralizing from Bangkok, and reduction of rural-urban migration (primarily to Bangkok). Indonesia's regional concern is with rural overpopulation and unemployment in Java and Bali, which cannot be relieved by further productivity gains in what is already a highly intensive and productive agriculture. Indonesia is also faced with large increases in the labor force for the next decade, consisting mainly of poorly educated and unskilled workers. Migration to new lands is a limited partial solution, but creation of a large number of rural off-farm jobs is essential to contain the growth in unemployment and avert massive migration to the cities.

It is a mistake to stress the need to train the nearly two million Indonesian workers entering the labor force every year, or the many millions already in it who have little skill and are underemployed. What will they do? A supply of skills does not generate demand for them.

The stress on identifying, training, and assisting entrepreneurs is one facet of a strategy of increasing demand. But ultimately entrepreneurs identify potential demands, and actualize latent demand. They are essential, but they are not sufficient. Increasing demand through import substitution does not appear in the cards for the near-term future. The easy phase of import substitution is over. What import substitution opportunities remain require the very resources and capabilities that are lacking in Indonesia. This is one reason why foreign investment outside of natural resources is lagging. Export promotion offers somewhat better prospects. But the most likely prospects are in natural resources, whose exploitation employs few workers (oil being the extreme case). The demand originating from such exports accrues largely to the government and its enterprises, and must be recirculated if it is to contribute to employment-generating domestic demand.

Export of the products of labor-intensive industries (other than specialized agricultural products) is not a promising prospect for expanding demand in the near-term future either. Indonesia must compete in world markets with other nations with almost inexhaustible supplies of low-skilled labor that is lower-cost than Indonesia's. Some countries also have more abundant supplies of cheaper skilled labor. Indonesia would have to increase the productivity of its low-skilled labor before hoping to expand employment through exports. Statements are often heard that in this or that industry wage costs are only 5 percent of total costs. Only in a few manufacturing industries is it true that direct payroll is a large enough share of total costs to influence location in low-wage countries. Modest variations in the efficiency with which plant and equipment are used can be far more important in locational decisions than even very large differences in wages. Labor-intensive products that take advantage of Indonesia's natural resources also offer the best prospect for export growth as a source of increased demand for workers. Policies designed to lower the cost of the business environment will both attract more foreign investment and permit the export of products whose cost of production at present is too high. And the earnings from oil, gas, minerals, lumber, and fisheries must be widely distributed if they are to provide the demand that permits entrepreneurs to survive and grow, and employment to rise.

Investment in advanced technology at this stage of Indonesia's development is a brake on growth and a drag on employment; it concentrates too large a proportion of scarce technical and professional manpower in sectors that are capital-intensive, depriving the rest of the economy of skills, management, and capital. Capital-intensive industries are great foreign exchange consumers, potentially reducing net foreign exchange earnings and derived demand available for the goods and services that can be produced with Indonesia's current factor supplies and that could provide large numbers of jobs.

Indonesia is at a difficult crossroads. No one policy direction promises to generate large numbers of additional jobs. Import substitution is least promising; export promotion in selected areas offers better prospects. But careful allocation of investment and foreign exchange resources as well as scarce human capital (distributive policies) are needed to generate adequate growth of demand, which to be implemented requires formation of large numbers of small- and medium-size enterprises, with the credit, technical assistance, and managerial training to support them.

In considering how Japan and the United States might increase technology transfer to LDCs and improve the efficiency of the transfer process, it must be kept in mind that any policies or programs must have the consent of the recipient nation. Now that foreign aid is predominantly on a loan basis (even though, in the case of AID and JICA, the loan terms incorporate a substantial element of subsidy), the donor (now lender) lacks the leverage previously enjoyed when aid was in some respects a free good. Recipient nations tend to behave as though they were paying full cost. Technology transfer by private firms was never regarded as a free good, even in the case of direct investment. Hence recipient countries have always exercised some discretion in accepting such investments.

Indonesia: Management as the Critical Shortage in Absorptive Capacity

We have concluded that the key limitation to expanded technology transfer to Indonesia is an inadequate absorptive capacity. This has to be increased in order to reduce the costs of investment, training, business operation, and associated technology transfer. U.S. investment in particular is too low because the costs of the business environment are high. A large increase in the number of competent managers and administrators at all levels (specifically at lower and middle levels) of the government (and particularly in the regulatory agencies) is a necessary condition for reducing the uncertainty and the anticipated costs of business entry and operation. However, it is not a sufficient condition—there must be determination and persistence at the top—because deeply engrained attitudes and practices are not quickly changed. But no ambitious reform will succeed without the human infrastructure to implement it.

The top three priorities for expanding absorptive capacity are management, management, and management. Foreign firms need better-prepared management trainees if they are to run their operations efficiently and if they are to reduce the number of expatriates. Domestic firms need a much larger supply of competent managers if they are to invest, modernize, expand, and generate demand for managers who will induce turnover from foreign firms and the associated technology transfer. Turnover of trained and experienced managerial and professional personnel is limited by an inadequate formation rate of domestic firms able to use their abilities. Government needs many more trained and experienced managers. These managers are needed to operate the many public firms that play a major role in the economy, to improve the efficiency and reduce the costs of government administration and regulatory services to business, and to improve the collection and dissemination of information.

Finally, if Indonesia is to cope with the combined problems of densely populated rural areas with low income and high unemployment and underemployment (and the large growth in the labor force in the years ahead), it must generate millions of jobs in many thousands of mainly new and small businesses scattered through the many areas with large supplies of unemployed and underemployed workers, most of whom have limited education and skills. These firms need minimal management skills, and need access to well-trained and experienced advisors on management and other problems if they are to contribute significantly toward moderating Indonesia's central problem for the next decades.

What is needed is large numbers of managers at lower and middle levels who have developed a systematic analytical way of thinking about their organization and their work, rather than who have acquired specific skills. In fact, mastery of the former will indicate which specific skills are most needed, and how they can be put to use in the context of the organization with its needs, resources, and constraints. Individuals who may have mastered a systematic analytical approach, and may have been trained in specific skills useful in the context of large complex organizations in advanced countries, need to adapt their approach and their skills to situations where prompt and accurate flows of information are lacking, and where the support staff and facilities for the exercise of these skills are deficient.

Business Contribution to Management Training

U.S. firms and educational institutions have a comparative advantage in management training. This is due in part to formal management education, in part to the high mobility of management in the U.S. economy. Thus, general management skills are stressed rather than nontransferable firm-specific skills.

Formal, structured training for managers, with stress on lectures, tapes, demonstrations, and reading materials—with all its limitations—has two advantages over more informal, on-the-job training. First, it is more easily expanded to accommodate more trainees, and second, typically its contents are more readily transferable to other firms and even from one industry to another. It is less firm- and industry-specific.

Another reason for the U.S. advantage in managerial and professional training is the greater willingness of U.S. firms to promote Indonesians to senior positions, and the firms' willingness to advance able workers rapidly, compared to the Japanese emphasis on seniority and lockstep advancement. U.S. firms seem to concentrate more on training Indonesians for higher management positions simply because they provide these opportunities more readily or, conversely, because they are less able and willing to maintain a large staff of expatriates.

Banking and Finance

The banking and finance industry offers a special opportunity to apply its training capabilities to Indonesia's problems, if these capabilities can be institutionalized and compensated. Leading foreign firms in this sector have large and well-structured training programs for managerial and professional staff, because they are expanding, and because of high turnover resulting from the expansion and modernization of the banking and finance industry in Indonesia. This industry is also an example of system-specific skills in a system that is economy-wide in application. Finance and accounting specialists are in demand in other industries as well.

In time turnover rates will drop, expansion will slow down, and some of these training programs will be reduced or discontinued in the host countries, and training capability diminished. This could be avoided by appropriate insti-tutionalization and compensation.

Much of the training capability of foreign banks, however, is not in Jakarta but in regional centers—Manila, Singapore, and Hong Kong. The large number of trainees who have completed their training and in many cases moved into domestic financial institutions, however, constitutes a reservoir of training ca-pability. And the role of MNEs is not just in staffing but in helping design, oversee, and evaluate domestic training programs. Related institutions in the United States, such as the American Bankers Association and the Banking Institute in New York, might assist in this task. U.S. banks, because of restriction to state branching (and in some states to unit banking), are hostages to the fortunes of a state or a smaller community. Hence, the role of their loan officers is more appropriate to the needs of the densely populated rural areas of Indonesia than that of the loan officer frequently rotated in a centrally run nation-wide bank.

The leading problem of the Indonesia economy for the next decade is generating large numbers of jobs, particularly off-farm jobs, to employ the millions of workers who will be entering the labor force in the years ahead (as well as more millions who are at least seasonally unemployed). Large numbers of new firms, small firms in particular, must be established, and scattered through-out the densely populated areas of the Indonesian archipelago. To do this, information on investment opportunities must be generated and disseminated; resources, credit in particular, must be made available for investment; and technical assistance and information for production, marketing, and related purposes must be made available.

The key resource to reach often remote areas is the local bank loan officer. Ideally the loan officer would not only be able to evaluate or have evaluated investment proposals for which credit is sought, but would also be an informa-tion broker on investment opportunities and feasibility studies; would seek out potential entrepreneurs; and would be a technical assistance broker who could give the entrepreneur access to necessary special assistance. The officer would also have to be an entrepreneur, because in most cases there is not a ready pool of local entrepreneurs ready to move on the signing of a note; he should aggressively promote local enterprise. The officer would have to oversee invest-ments for which bank loans are made, to assure that the loan proceeds are spent as intended, and to protect the bank from default by knowing when and how to offer assistance, because many loans would have no collateral. (The widespread practice of bank loans at negative real rates of interest with no follow-up oversight and high default rates is not conducive to a large volume of new enterprise and employment generation.) This is much to ask. And bank loan officers with these qualifications are needed by the thousands if they are to reach the millions. (For a similar recommendation see Young 1983.)

The foreign banking and finance community is well-positioned to train large numbers of bank loan officers should appropriate institutional arrangements and reimbursement be made. Ideally such trainees, beyond formal training and some on-the-job experience, would, once employed as bank loan officers, receive one or more review visits as an extension of on-the-job training. The banking and finance industry might be induced to train well in excess of its own needs for years to come, through cost reimbursement, and by the expectation of widening the market for its services.

A separate institution would be more appropriate than expanded activities of existing banks, both because of the high compensation otherwise required for trainees, and because the training for loan officers to serve in often remote areas has to be somewhat different from that for international banks with activities concentrated in the major cities and with large borrowers. Also, the training of the larger foreign banks is conducted largely in regional centers outside Indonesia, and in English, whereas training for local bank loan officers would have to be in Indonesia, and in Indonesian.

Such a training institution need not be associated with any particular bank or with the financial institutions of any one country, and would be staffed by Indonesians with at least initial assistance of expatriates from several nations. There should be several training locations (drawing upon nearby banks and universities for trainees) and located in areas of need and opportunity for job creation through formation of new firms, introduction of new products, and processing activities. Training should be in regions where such bank loan officers are most needed, and should be geared to trainees living in the region. Training in the major cities, cheek by jowl with large firms and international banks, might divert the better trainees to them. A further advantage is that training can be somewhat differentiated to reflect the special needs and opportunities of a region in which it is conducted, and in which the trainees are expected to work.

The loan officers required are unusual in that they would have to function effectively without the ready access to special expertise available in the major cities. They would have to be well versed in project evaluation, as many promising borrowers would have no collateral. They should be well paid, in accordance with their qualifications, in order to assure their employment in areas where they are most needed, and that they remain long enough to become truly effective. It is probably not practical to require a certain number of years of approved employment in areas of need in return for the free training and salary while in training.

One component of the program of instruction should be provision of information on the availability of inventories of technology and sources of technical assistance, such as VITA, IESC, Project Sustain (food industries), or Appropriate Technology. These resources are underutilized in part because of lack of information on their availability (their cost is also a deterrent).

Three private management training institutes have related programs on a modest scale. LPPM (Lembaga Pendidikan Kan Pembinaan Manajement—Institute of Management Development and Education) has a one-year course for training small business consultants. These consultants are intended to be supervisors of bank loan officers, advisers in government agency development offices, and managers of large commercial companies seeking to penetrate new markets and of industrial companies developing subcontractors. Prasetya Mulia has a training program for small business entrepreneurs. Its associated "extension" work includes facilitating credit for small businesspeople. A third training group, the Indonesian Institute for Management Development (established in September 1984), offers short programs for business executives, as well as an MBA degree program.

The Construction Industry as a Source of Enterprise

The discussion of the role that could be played by local bank loan officers assumed an available supply of local entrepreneurs. In fact this supply may have to be discovered and developed.

The industry in which the most training has been conducted by U.S. and Japanese-related firms (engineering consulting firms, construction contractors, and oil firms) and which is likely to continue as the leading industry for training, is construction. Sectors include oil and gas production, transportation, refining facilities, turnkey plants for other public firms, and major public works projects. Because much of the hiring for construction projects is done in the region where they are located, large numbers of workers are trained, and many of them are out of a job on completion of the project cycle. Thus although there are localized shortages of skilled construction workers, there are surpluses in other areas. In fact Indonesia is exporting a substantial number of construction workers for temporary work in the Middle East, and attempting to increase the number working there. Thus in construction there are both a large training capacity and experience in the U.S. and Japanese firms, but also a considerable stock of experienced workers.

It has been suggested that experienced construction workers constitute an easily and quickly trained pool of workers for other skilled occupations—they are much more easily and quickly trained than unskilled and inexperienced workers (Helfgott 1973, 239). Perhaps more important, a large number of potential entrepreneurs and new small businessmen could be found in this pool of skilled construction workers. Professor Dorodjatan of LPEM (Lembaga Pendidikan Ekonomi Manajemen) University of Indonesia, in a study of four communities, found that typically it takes 15 years for a new unskilled construction worker to advance in experience and responsibility to the status of an independent foreman-contractor. The lack of rigid or vigorously enforced construction codes allows opportunity for ability and enterprise to develop quickly, which would be constrained by too much regulation of the industry. In the years ahead, large numbers of construction workers should be in such a position, given the rapid expansion of public works and resource industry construction in the 1970s. These independent foremen-contractors, who already are a significant factor in the construction industry, are risk-takers; the very project cycle nature of their industry is a selective process for risk-takers. They have experience in management: they are familiar with organization of production processes, and allocation and supervision of workers. And they are to be found in many locations throughout Indonesia, and not just in large cities, as is the case with other skills and managerial and entrepreneurial experience.

The local branch bank loan officer, properly trained, should be on the lookout for potential entrepreneurs to start new small businesses, or to produce new products. The loan officer should have a portfolio of investment opportunities. The construction workers with long experience, including management experience as foremen, and entrepreneurial experience as independent contractors, are likely prospects.

What can be contributed by foreign firms that train such workers and give them experience in the process of building oil and gas facilities and public works? They can provide information on workers at the end of a project cycle, or on former workers with long experience whom they judge to have entrepreneurial and managerial talent and inclination. This information would be provided to banks and other institutions interested in promoting new and small business, as well as to prospective entrepreneurs among their employees and subcontractors. Construction companies might consider, as part of the phasing out of their projects, some follow-on orientation and training for foremen and small independent contractors as potential entrepreneurs.

What needs stressing is that, in Indonesia, increasing the supply or improving the quality of skills is not enough; this concentrates on the supply side, and neglects demand. The discussion above of the role of loan officers and the potential for enterprise among experienced construction workers is aimed at generating additional demand, without which the investment in skills is wasted.

What is suggested is an industrial-service extension system, with the bank loan officer roughly occupying the position of the county agent plus rural capital supplier, able to draw upon a wider range of expertise than that required of an agricultural extension agent. A further difference is critical: the entrepreneurial function. The county agent does not need to identify potential farmers and encourage them to go into farming. The agent's market is ready-made, and the agent only intermediates between the needs and problems of the farmer and market and technical information sources. The credit officer must help create a market as well as service it.

Training for Business and Industry

Institutionalizing the Training Capabilities of Foreign Firms

How may training capabilities of U.S.- and Japan-related firms be used more effectively to meet the needs of the Indonesian economy for skilled workers? How may they be divorced from investment cycles and turnover experience of U.S. firms? The first consideration is that training is expensive: trainees must be compensated while training, trainers are scarce and expensive, and space and equipment may be needed. A profit-seeking firm would not normally train workers in excess of its anticipated needs. Nor would it maintain a training program once its needs fall below some critical level beyond which other alternatives become cheaper than its own training: hiring experienced workers, or paying to have new hires trained elsewhere. The other side of the coin is that fluctuating or discontinuous training activities may themselves be expensive, and a steadier flow may offer substantial cost advantages.

Training provided by business firms for outsiders who have little prospect of obtaining employment in the firm conducting the training and who are not paid while undergoing training is neither viable nor useful. Even though training programs that compensate their trainees can attract trainees without a clear prospect of subsequent employment, these programs are obviously costly. Training programs that do not compensate trainees must offer a clear prospect of subsequent employment on successful completion of training. If a firm is to train more workers than it needs, then other firms using the same skills must be prepared to offer some assurance of subsequent employment.

For other firms to offer assurance of subsequent employment, there must be an appropriate organization or institution: an industry association, or a skill/craft association, for skills that are not industry-specific. Furthermore, training would stress those components of skill that are industry-wide or occupation-wide, and tend to neglect those aspects of work skills that are firm-specific, as these would be useless for trainees subsequently employed by other firms.

An industry association or skill/occupational association cannot be limited to Japanese- and U.S.-related firms, but should encompass all firms interested in hiring trainees in particular training programs. What U.S.- or Japanese-related firms can do is take the lead in sponsoring training programs; others will have to join in providing some assurance of employment upon successful completion of training.

The question remains, why should Japanese- and U.S.-related firms undertake such programs? What is in it for them? There are indirect benefits: public relations, and increased supply and hence lower cost of skilled labor (whose benefits however mainly accrue to other firms). There could be tradeoffs: a more lenient attitude toward employment of expatriates, much of whose contribution may consist of training; more favorable treatment in terms of tax liabilities, import licenses, credit, and the like. To the extent that the firm is either expanding, or facing high turnover, or both, training more workers than it needs in the immediate future can make sense from a longer-run perspective: they constitute a pool of pretrained potential employees from which to select new hires. Yet all these considerations together will not induce training activities much in excess of a firm's perceived needs. Some compensation for excess costs is required.

Two approaches can be employed. One of these is training by a particular firm of workers in excess of its own needs, with excess trainees obtaining employment in other firms. Goodyear pioneered this approach. (In the case of Daimler-Benz, trainees are sent by other firms, who pay the training costs including salary for the trainees.) Not all firms can consider this approach. Some firms employ skills that are too firm-specific. But many firms are too small to maintain a cost-effective training program, and could be customers for such a program. Yet firms in a position to conduct effective training of workers in excess of their needs must have some inducement. The public relations value of such a program is real but limited. In the case of West German firms, the Federal Republic of Germany compensates them in the amount of 200 to 400 deutsche marks per trainee per month, provided they train workers in excess of their needs. For Daimler-Benz this is less than 50 percent of training costs. But because it applies to all trainees, including those trained for the company's own needs, it is a powerful inducement. One issue is the determination of how many trainees are in excess of the training company's requirements, as it trains not only for itself but also for associated Indonesian companies. It also trains not just for its current needs but in anticipation of turnover and the possibility of future expansion. Because the apprenticeship program lasts two years, anticipated needs are an important consideration.

Government subsidies for company training of unskilled workers beyond its own anticipated needs have been used in the United States; it is an approach to be considered. Nor of course need inducements be restricted to AID or OPIC. The GOI might itself offer a variety of inducements as appropriate.

Another method of compensating the training firm for costs in excess of those needed to meet its own needs is to have firms hiring "excess" trainees reimburse the training firms for its excess costs. This might be a mutually profitable arrangement, particularly for smaller firms that might find an internal training program very costly, but it is only likely to be feasible for industry-specific skills in industries with effective associations that can police these transactions. If a firm is free to hire "excess" trainees without reimbursing the training firm for its extra costs, then the reimbursement approach may break down. This is an example of the free rider problem.

A second institutional approach is to set up an autonomous training institution supplying scarce skills to an entire industry, as exemplified by Toyota-Astra and National Gobel. Such an institution has greater flexibility than one which is an integral part of an operating firm. It may be able to attain economies of scale beyond the reach of individual firms. And it can have greater flexibility in adapting its training activities to the needs of an industry rather than those particular to a firm. It would not face a choice between firm-specific training on one hand and occupation-, industry-, and system-specific training on the other. For Japanese firms, whose training programs have a much larger firm-specific component than is characteristic of training by U.S. (and by other foreign) firms, this approach may be preferred to the alternative of training of "excess" workers by particular firms for other employers. An autonomous training institution may also be in a better position (particularly insofar as it monopolizes industry-specific training) to cover its costs by contributions from firms who hire trainees. For one thing, there is no issue as to what percentage of costs should be borne by the training firm, and what by employers, if in fact the training firm (foundation) does not hire any trainees. For another, there is a more even-handed competition in the labor market. (This does not mean that the free rider problem can be entirely avoided, but its incidence could be minimized.) It should also be easier for governments including Japan, the United States, and GOI, to subsidize such an activity if appropriate, because it need not be associated with firms of any particular nation, and ideally would be multinational in staffing and support.

The present situation in Indonesia is that government training and vocational programs are inadequate in quality as well as in quantity, and that they are poorly related to the needs of business. The problem of quality probably cannot be solved by government without the close cooperation of industry. Government simply does not pay enough to attract and hold good instructors. In skills that are new to the country, this is inevitable; foreign firms are likely to have a monopoly of capable instructors, and will certainly pay much more than the government will. For the next decade at least, Indonesia must rely on business firms to provide a large share of the scarce skills needed for economic growth and development.

Issues that need to be considered, whatever institutional form is adopted or whatever inducements are used, include: what skills are appropriate for "excess workers"? How are trainees selected and by whom? How shall trainees be compensated? Presumably the firm doing the training would use the same procedures it employs in selecting trainees for its own needs.

Finally, what happens to the trainees in excess of the training firm's needs on successful completion of training? If skills are carefully selected for scarcity, trainees will find jobs on their own, and prospective employers will come to the training facility in search of them. Where skills are not acutely short (and there is a case for training even here, in terms of moderating wages and turnover costs), or where demand for them is in areas far removed from the training site, some institutional arrangement is needed to match prospective employers with trainees. Nothing could kill off a well-intended training program for "excess" workers than the discovery that these workers were in "surplus", that is, unable to find jobs related to their training.

The discussion of management training needs, suggestions for contribution by foreign firms, and the institutionalization of training for business and industry was in the Indonesian context. Although Thailand could also benefit from similar approaches, its needs are less acute; its educational and training systems are better than Indonesia's and more closely related to the needs of the economy. Much of the institutionalization of training suggested (and that has been noted in Indonesia) is a response to the acute inadequacies of the Indonesian educational and training systems, and could be viewed as a temporary expedient. In Thailand it is an expedient whose time has past for many skills, but which has a place in growth of new industries demanding skills new to the economy. The alternatives of expanding the training programs of particular firms, business foundations, and institutes, or developing the capabilities of national educational and training systems depend on (1) the numbers that need to be trained; (2) the number of prospective employers; and (3) the life expectancy of the skill. If the need for training is expected to be short-lived, or if the skill is too firm-specific, then it does not justify institution-building or curriculum revision.

The examples given above of institutionalization of training activities or capabilities of MNEs refer to country-by-country activities. Whether or not some of them can be opened to trainees from other countries depends in part on the language of instruction, which is the country's own language for production skills, but is usually English for managerial and professional skills. A substantial part of the training for high-level skills is conducted in Asian regional centers by U.S. MNEs, in Japan by Japanese firms. The possibility that some of them might be opened to employees of other firms or government agencies, and the conditions under which they might be opened, are worth investigation.

Educational Technology

Given the great scarcity of highly educated people in fields relevant to technology transfer, and the high cost of expatriates (and their limited utility in situations where training must be in Indonesian, and/or where it must be adapted to local needs and conditions), how does one make the best use of these limited resources? With a scarcity of the relevant skills, and the better pay, living conditions, and opportunities in large cities, it is inevitable that most of the critical human resources should be concentrated in large cities and particularly in Java.

Ways of imparting information that do not rely on face-to-face contact, and that are not interactive, have the advantage that they can reach large numbers of trainees even in remote regions. The presentations and demonstrations of the best instructors are packaged in videotapes, film, radio, and to some extent in reading materials, and in these forms can be made available in many and even in remote locations. Properly used, in institutional settings, such techniques need not be inferior to the labor-intensive face-to-face interaction of trainer and trainee; they are extensively used by some of the larger U.S. firms in Indonesia for their own employees and for their customers. There are infrastructure constraints to be considered: most villages have no electricity.

Dean Jamison (Jamison 1975, 47), in the somewhat different context of secondary school education, advocates discontinuing the costly building of additional schools and training of additional teachers and resorting instead to radio correspondence instruction. The cost of radio is only one-fifth the cost of TV. In the context of training bank loan officers, and new and small entrepreneurs, modern communications technology could vastly extend the reach of a very limited supply of capable instructors and advisers. What foreign organizations can contribute is content—whether adaptation of teaching materials already developed by foreign firms or private nonprofits, or assistance in developing suitable materials that could then be disseminated by means other than, or in addition to, face-to-face instruction.

Thailand: Building a Technological Infrastructure

Manufactures already account for over a third of Thailand's exports. Although manufactured exports are mainly textiles and processed agricultural products, prospects look good for exports in other industries. Export-quality production, as well as production at internationally competitive costs, requires development or strengthening of the technological infrastructure needed by the industries expected to locate or to expand in Thailand. Part of this technological infrastructure is capabilities (institutional and individual) for applied and adaptive research, to be discussed later.

Apart from the educational system to supply needed human resources and to help scientists, engineers, and managers to keep abreast of latest developments, a technological infrastructure for technologically sophisticated industries must have the following capabilities (U.S. National Bureau of Standards 1977, 4):

1. Standards: a conceptual system defining relevant phenomena, quantities, and units
2. Basic technical infrastructure: a documentary specification system of physical standards, techniques, and instruments, supported by a reference data system, reference materials, and an instrumentation industry (which initially would rely mainly on imports)
3. Realized measurement capabilities

4. Dissemination and enforcement network, consisting of central and regional standards authorities responsible for physical measurements, standards and testing, and calibration laboratories and services, and regulatory agencies that specify requirements

5. End-use measurements: quality control at the production end

Certain scientific and technological areas have been singled out for development. These include biotechnology, electronics, nonferrous metal technology, resource use and management, and selected chemicals.

The U.S. National Bureau of Standards has ranked industries in terms of measurement intensity, defined as the percentage of value of product consisting of measurements. Electronic components is a close second to aircraft manufacturing. Office equipment, computers, and radio, TV, and communications equipment are nearly as high, followed by plastics and chemicals. Clearly Thailand's intended direction of structural change calls for expansion of this aspect of technological infrastructure. Nor are needs limited to industry. Agriculture has extensive needs for humidity measurement as a basis for efficient humidity control. In a humid tropical climate such control, and associated containment of aflatoxin, is essential to maintain the quality of products, and needs to be done in a cost-effective manner. In chemicals, especially hydrocarbons, the critical measurements are of process control parameters. Some of the institutional framework at the national level has been in place for years. These are a Ministry of Science, Technology, and Energy; a National Research Council serving as an advisory body; centers for national documentation, for instrument repair and calibration; the Center for Thai National Standards Specifications. There are a large number of research institutes in a variety of fields, particularly in agriculture. The Thai Development Institute, modelled after Korea's KDI, is in process of formation. But there are problems of staffing, management, funding, and particularly of implementation.

Large MNEs in advanced technologies typically have some of their own in-house capabilities. But to attract smaller firms, and to encourage the development of domestic firms in these new industries, technical services and facilities should be available. Because these are diverse, and demanding in their human and capital requirements, it is important to concentrate on a limited range of industries and associated technological infrastructure rather than attempt to develop across the board.

Both U.S. and Japanese government agencies could be very helpful with technical assistance in developing and upgrading institutional capabilities. An agreement on cooperation in science and technology was signed between Thailand and the United States in 1984. It is intended to facilitate cooperation not only between government agencies but also universities, research centers, and private-sector institutions and firms (Briskey 1984). In some industries, major firms or industry institutes in fact set industrial standards and may help monitor them. At the final stage of quality control, Japanese industry is widely regarded as the world leader.

Strengthening Research and Development

Most kinds of research are best left to the advanced nations, which have the resources and the pressing need for research, and whose research findings will become available everywhere sooner or later, and much of it at no cost to LDCs. Research results only available at substantial cost or with considerable delay in general are not those suited for immediate implementation in LDCs. In one area, however, some of the needs for research are too region-specific for an LDC to rely on research done elsewhere, and too pressing to await the time-consuming process of transfer and diffusion of research done elsewhere. This is the case of research in agriculture and on related renewable resources, such as forestry and fisheries and, more broadly conceived, on local ecology.

Agricultural Research and Development

Advanced nations may do little research on crops important in tropical areas, or potentially important in them. Soil types are different, and vary from region to region. Some soils are so acid that the standard usage of nitrogen fertilizers may be harmful, and practices may have to be modified and carefully monitored. High temperatures the year around accelerate the process of organic decay, and heavy rainfall leaches the soil of both organic matter and soluble minerals. The rain washes out fertilizers and pesticides more rapidly than in temperate zones. The shift from species diversity and dispersal in natural tropical vegetation to the monoculture of commercial farming, as well as the development and widespread use of hybrid seed, makes agriculture highly vulnerable to insect pests and plant diseases. Insect pests and plant diseases are a year-round threat, and may mutate and develop resistance to pesticides and herbicides more rapidly than in temperate zones (Janzen 1973).

Proposed research institutes must be able to respond quickly to localized problems. They must develop crops, seed, and cultural practices best suited to local conditions and resources, and have the appropriate network of information flow in both directions. They must respond appropriately to the needs, problems, and possibilities of the producer, and to crop and cultural practice recommendations from the research institute to producers. Research institutes must adjust their priorities to the needs and problems of local agriculture so that they can provide leadership as well as respond quickly.

The problems of tropical heat, poor tropical soils, excess or untimely rainfall, of plant diseases and insect pests, are not the only, or perhaps in the long run, the main concern toward which agricultural research institutes should address their efforts. With the growth of population, the extension of the agricultural land area, the commercialization of agriculture, often the lack of secure tenure, as well as the lack of knowledge of consequences, agricultural land may be exploited and abused, reducing its future productivity and causing large negative externalities. Erosion can silt up irrigation canals and rivers; improper use of fertilizers and pesticides can contaminate downstream water; and improper use of irrigation can increase the salt content of water to the point that it becomes harmful if used for irrigation downstream. Some agricultural research institutes at least must see the rural-agricultural environment whole, in all its interrelations. They must have an ecological perspective within which to identify and assess problems and to evaluate changes in crops and cultural practices.

Some of these special needs of LDCs can be met by international organizations, such as the IRRI in Los Baños. Others call for capability within a nation, and specific areas within the nation that have distinctive problems or resources. The need for close linkage between research, training, and extension services further underlines requirements for developing indigenous capabilities.

Thailand.   Thai agriculture uses little fertilizer, pesticides, or herbicides. Until recently it increased output by expanding the cultivated area. But this approach has now about reached its limits, and henceforth Thailand must concentrate on increasing yields, which are quite low, not only of rice but of other major crops. This means a large increase in the use of fertilizers and pesticides, and the use of new seed varieties with different sensitivities and demands. These in turn will call for an intensification of the work of agricultural extension agents, and of applied research on a regional basis that will supply extension workers (and through them, the farmers) with the knowledge to sustain and increase yields. The new rice hybrids in particular impose needs for organization and coordination because of the requirement for synchronous planting, for credit to purchase needed inputs, and increased peak load needs for water, labor, inputs, crop storage, and transportation (Goodell 1984). Only 25 percent of rice production consists of new hybrid varieties. Only a quarter of rice is under irrigation, a share that could be doubled.

New crops are needed to increase the incomes of the Northeast and other poor areas in Thailand and to replace crops whose export markets have been reduced (cassava). Such crops require applied research with a regional focus and appropriately instructed extension workers, and in some cases, the development of appropriate infrastructure. An increase in the proportion of agricultural output marketed, and an increase in exports, require expansion of appropriate infrastructure: transportation, storage, processing, inspection, and quality control. Much of this would be new private business, hence the stress on management training previously discussed.

Exports of agricultural products are handicapped by the lack of testing facilities to verify the quality of agricultural products, of fishery products, and of canned and frozen products for export. Canned fruit and juices face a growing market but require promotion. U.S. and Japanese food products firms, and agricultural agencies, intermediated by JICA and AID, can make a major contribution at several stages. Facilities for testing, and such training as may be required for them, are the last stage. Quality control at earlier stages needs to be improved. The storage and transportation system needs improvement at every stage from the farm or fishery to final processing or sale. Aflatoxin is a major problem for field crops. Technical assistance can play a role here. Quality control problems occur also at the crop-growing stage and must be dealt with through extension workers equipped with knowledge supplied and to some extent developed by regional agricultural research facilities. The role of the United States and Japan here is to provide information, technical assistance, and research support.

U.S. MNEs in the food industry are playing a modest part in modernizing Thai agriculture. They purchase crops from contract farmers and provide them with technical assistance (a private extension service), credit, and other resources they may need to produce a product of the quality desired. They also contract with local universities for research and provide export markets. But U.S. MNEs are not allowed to own land or to produce except as minority partners with a Thai joint venture (production of seed is an exception). They also provide export markets. Such firms, of whatever nationality, should be encouraged to expand their activities.

The primary need is not for qualified personnel. Kasetsart and other universities have good training programs at the graduate as well as undergraduate levels. Some Thais need to be sent abroad for advanced training in selected specialties. Some foreign consultants should be sent to Thailand, but the primary problem is better utilization and systematic development of the human resources already on hand. This has been described as a research management problem. There are over 80 research centers, with little coordination and inadequate support. Most of the capability is in the universities, whereas most of the funds come from the Ministry of Agriculture. Each department within the ministry has its own research facilities (and some their independent extension services), in many cases with so little interaction with farmers that the Extension Department has established experiment stations of its own. Some universities have their own dissemination and extension programs. The universities compete for research funds from the Budget Bureau through the National Research Council. Funds are granted on a one-year basis. The lack of continuity in research funding hampers both research and the development of the capabilities of researchers.

There appears to be no shortage of extension workers. One respondent claims there are too many—that they pester farmers. But many are poorly qualified, with only agricultural vocational school certificates. It is impossible to retain university agricultural graduates as extension workers. Their training needs to be improved if they are to become effective translators of research results into practical advice, and conveyors of information on farmer needs and of problems to the research laboratories.

The lack of coordination, the duplication of facilities and efforts, are well recognized, reflecting traditional Thai attitudes toward the management of power. What the U.S. and Japanese governments could do is make research management consultants available if desired. What they should do is provide financial support for university research, which they, and some nonprofits are already doing, on a larger and continuing basis. In particular they should help strengthen the research capabilities in the regional universities in more remote parts of Thailand, contributing incidentally to the quality of training and retention of qualified specialists. Local university complaints are that they cannot work directly with the U.S. Department of Agriculture, and that it is hard to initiate projects from the local AID office.

Japan and the United States are already cooperating to strengthen the research capabilities of Khon Kaen University in the Northeast, which should help correct the underemphasis on rain-fed agricultural research. Two areas in which they could provide needed technical and financial support are forestry and fisheries.

Thailand's main need for forestry research at this time is for the protection of agriculture. Supplies of commercially valuable timber are nearly exhausted, and cannot be replenished for a long time, particularly teak, which takes many decades to mature. Research is needed on the consequences of deforestation and how soil erosion and the siltation of water supply systems may be minimized. In the case of coastal swamplands, the cutting of mangrove threatens the nurseries of fish and shrimp. More information is needed on the state of affairs, on the damage being done, and on effective countermeasures.

Another need for forestry research is the selection or development of fast-growing trees to be used in village woodlots, supplying fuel, small construction timbers, and food as well. This plan will safeguard those forests that remain from destruction for firewood, the principal source of fuel for cooking and other household purposes. It is also necessary to obtain full cooperation at the village level from the very initiation of any village woodlot project, to assure that the trees will be allowed to grow to the proper harvesting size and that they will be shared among villagers in an acceptable way. This is an extension as well as an applied research task.

In terms of fisheries, there is little knowledge of fishery stocks. There is need for information and research on the Gulf of Thailand, which has been overfished, and polluted in part. Breeding areas for fish and shellfish, especially shrimp, are being destroyed by cutting of mangrove. Information is needed on the present status of aquatic life within Thailand's territorial waters; research is needed on the possibilities for replenishing offshore fisheries and coastal breeding areas. Improved capability for inspection of catch and quality control of processed and frozen fish and seafood are needed. Aquaculture raises a different set of problems. Siltation (as a result of deforestation) and the effects of fertilizer, pesticides, and herbicides on fish must be studied and controlled.

Indonesia. The importance of increasing the incomes of farmers, whether on the farm or in off-farm employment, cannot be overstressed (Oshima 1984). Such a policy would reduce the migration of rural farm population to urban centers, thus reducing the need to concentrate infrastructure investment in cities in response to the needs of a large influx of poor, unskilled, and largely unemployed people. An increase in farmer incomes would also expand regional demand to support production of a wider range of goods and services with local and regional markets and widen the national market for other industries (Stewart & Lee 1986).

In the most densely populated areas of Java and Bali, the most promising (or least unlikely) prospect is increasing off-farm income, whereas elsewhere increasing farm income is a more realistic possibility. The former is a supply-led strategy; off-farm production generates jobs and incomes that help purchase its output. The latter is a demand-led strategy: Higher farm output and income increase demand for both national and locally produced goods and services. Improving agricultural productivity, the introduction of new crops and new processing facilities, can be assisted by Japan and the United States through training and technical assistance for agricultural research and training organizations. This can also be done through investment in and assistance to agrobusinesses, and in some cases, by opening up markets to Indonesian agricultural exports.

Generation of off-farm employment in densely populated areas is a need to which Japan and the United States can contribute only indirectly. Nearly all the investment and new firms will have to be domestic. The training capabilities of the foreign banking industry and the experience of the foreign engineering construction industry in which U.S. firms are heavily represented might be tapped to stimulate development of new firms and products in the areas with very high rural population density.

Introduction of new hybrid rice strains in Java reduced employment of women harvesters by two million. The effects on aggregate employment of increased rice production are uncertain (Mears 1984, 134). The use of fertilizers is long established, and yields are over 4 tons per hectare. Additional gains are likely to prove expensive and to involve organizational changes in the use of pesticides and fertilizers. However, rice research and extension is relatively well developed. The potential for increased output and employment is better in the less densely populated islands, and in secondary crops rather than in rice. Most of the new lands in the outer islands being brought or to be brought into cultivation are not suitable for rice.

Although agricultural education has been much influenced by U.S. universities, the three functions of education, research, and extension are fragmented.

Whereas Bogor and Gadjah Mada, under the Ministry of Education, have good graduate programs in agricultural specialties, the research strength is in the Ministry of Agriculture, whose research institutes are reasonably well linked with the extension services. Research conducted at the universities is largely on a private "moonlighting" basis by faculty rather than an integral part of university activities. Hence support for applied research should be concentrated in the government research centers, not necessarily excluding some support for universities as well.

Support should be concentrated on strengthening capabilities in the outer islands, which offer the best prospects for increasing agricultural output and income. Support should also be concentrated on the important areas where Indonesia's research centers are weak. This means not only in rice (which is relatively well studied) or in the estate and industrial crops (research in which is conducted by private firms as well as by government). But support must also be directed toward secondary crops, forestry, and fisheries, all of which are weakly encouraged but all of which offer good prospects for growth. The greatest need is still for advanced training (some of it overseas) and technical assistance. Financial and technical assistance for priority research projects is also needed, but Indonesia remains short of the trained individuals who can conduct them. Here as in many other areas, management is a critical shortcoming, and improvement is a necessary condition to close what has been called "a yawning gap between planning and implementation" (Tarrant 1983, 6).

The needs for forestry research in Indonesia are diverse. First there is need for information on the extent and rate of deforestation and its consequences for the environment. Second, research is needed on appropriate harvesting practices for a sustained yield outcome with due consideration for environmental impacts. Third, research is needed on the best ways of reforestation, not only in terms of costs and minimal environmental damage but in terms of maximizing yields. Given the diversity of soils and climates in Indonesia, practices must be adjusted to each distinct environment. Fourth, research is needed on plant diseases and insect pests and how they may be controlled. Fifth, some attention should be given to the consequences of forest clearing for agricultural uses.

Forestry, and deforestation as a part of the transmigration program, are concerns of the outer islands. Densely populated and intensively cultivated Java and Bali have different problems. Population pressure has led to the clearing of steep slopes for agricultural use, with consequent erosion and siltation of irrigation systems. Cutting trees for fuelwood has further aggravated the problem, which calls for selective reforestation in the interests of preserving agricultural productivity. Research on multiuse fast-growing trees to supply villages in Java and Bali and spare forests has a lower priority than in Thailand. The availability of cheap kerosene lessens the demand for fuelwood. A fast-growing species is already widely used for fuel at the village level.

There are three forestry research institutes in Indonesia, but they are inadequate in number and quality. There are also problems of implementation, particularly in watershed protection. A Ministry of Forestry was established only in 1983. Its jurisdiction overlaps with that of the ministries of public works, internal affairs, and population and environment.

Fisheries is another priority area for Japanese and U.S. training and technical assistance. The Japanese in particular are assisting the fishing and fish processing industry. However, there is inadequate information to define sustainable catches and the consequences of offshore and onshore environmental changes on the potential for ocean fishing. Applied research capabilities are inadequate, given the very extensive and diverse ocean areas surrounding the Indonesian archipelago. Aquaculture is a relatively unexplored possibility, particularly in the extensive swamplands in outer islands, which are convertible either to agricultural use, or to aquaculture.

A problem in relating the work of experiment stations to the needs and problems of farmers is the fragmentation of extension services. Another is the multiplicity of government departments involved in rural-agricultural programs of an environmental, water management nature. In part for these reasons, only the better-led and organized villages are likely to obtain access to public services. This is where nonprofit organizations involved in rural community development can be of help, bridging the information and access gap to technical assistance, credit, and other resources.

Other Research

Research needs (other than those directed at agriculture, forestry, and fisheries, and at the processing industries based on agricultural products), fall in two categories: needs related to the welfare of the population, and needs related to export industries.

Both Indonesia and Thailand have needs for research in medical, nutritional, and public health conditions of the population. One might add family planning as another area. In Indonesia, inadequate nutrition is cited as a factor in low productivity. The primary need is for comprehensive, somewhat detailed, and up-to-date information, particularly in Indonesia with its far-flung borders and diverse local environments and living conditions. Research encompassing both the biological sciences and the social sciences is needed in order to cope with the health problems of the population. Diagnosing problems and specifying solutions is far from enough; people must be informed, motivated, and persuaded, or little is accomplished. Because polluted water is widespread and is a major factor in infant mortality as well as a medium of transmission for various diseases, adaptive research is needed to design workable and cost-effective systems to reduce the problem. This is an area where Japanese and U.S. firms can readily contribute technical assistance and financial support for selected areas of research. U.S. nonprofits have been instrumental in developing the human resources for applied research in the health field and for implementing the recommendations issuing from research.

Research aimed at the needs of export industries and new industries with export potential must be adaptive, very applied research. The problem is not always one of individual and industrial capabilities but of orientation. Scientists trained according to the Western model have a tendency to follow it without modification for the local conditions; they tend to be too theoretical, unconcerned, or unresponsive to the needs of industry and business. The status system in Thailand at least leads productive scientists to move from the laboratory to administrative positions. What foreign scientists can contribute in this case is example, but no more. Foreign organizations can reorient scientific efforts somewhat through financing adaptive, industry-specific research.

Thailand has selected certain fields for development of scientific and technological capability: biotechnology, electronics, nonferrous metals, and selected chemicals (particularly hydrocarbons).

Indonesia on the other hand has so many priorities that in effect it has none. At this stage in its development the creation of research capability in industry appears premature; lower costs through increasing absorptive capacity for production technologies and improved management should come first. The major exception may be mining techniques and research on containing the environmental consequences of mining and minerals processing activities.

The capabilities for consulting and technical assistance in industry-specific research are to be found mainly in U.S. and Japanese firms, though in some cases in industry research institutes, universities, and government agencies. Education for research personnel can only accomplish so much. It is a preparation for learning by doing, not a substitute for it. Foreign firms can contribute to developing local capability for research, consulting, technical assistance by offering the opportunity for local individuals and institutions (primarily the latter) to gain the experience for which they have the background, and in time to develop competence. In the long run it is cheaper than to claim local inadequacies as the basis for continued reliance on expatriates and foreign institutions.

One last aspect of research capability and effectiveness in relevant fields should be mentioned: the information networks to provide researchers with ready access to the work of others in other countries, or even in the home country. To some extent this means access to publications, although this alone is rarely satisfactory because of the lags involved, and because much new applied knowledge is not published. It also means institutional contacts, individual contacts, conferences, and exchange programs. Without them, research productivity suffers.

Feasibility Studies and Investment Promotion

Investment feasibility or prefeasibility studies to attract foreign investment to Indonesia and Thailand (and the indigenous capacity to conduct them) are inadequate, as is the outreach process to potential foreign investors. This is a much more serious shortcoming in the case of U.S. than of Japanese investors, because the Japanese trading firms (*sogoshoha*) gather information and disseminate it to smaller firms in Japan. To a degree, they also conduct their own investment feasibility studies. JICA has also financed many feasibility studies by Japanese personnel at the request of the Indonesian and Thai governments. U.S. firms are only beginning to develop similar capabilities. OPIC's contribution in this regard is limited and indirect, as it is a facilitator, not an investor.

Even if all major investing countries had the capability of the Japanese *sogoshosha*, it is inconceivable that the host country should not wish to have its own autonomous capability. It should develop the information channels to reach out to potential investors, licensors, and contract purchasers.

OPIC's approach toward feasibility studies is to have them done by the prospective investor, and to assist financially in this task. This is the ideal, but it does not deal with the problem of identifying and interesting potential investors. This "ideal" at least assumes some prefeasibility studies, with the objective of determining the areas in which feasibility studies should be conducted, and the industries/products in which to seek potential investors.

Thailand (and Indonesia) should not depend solely on investors who are already interested in producing specific products overseas, and are at the stage of deciding in which country to invest. They should also take the initiative in bringing investment opportunities to the attention of potential investors. To do this effectively they must have credible estimates of costs in all cases, and of markets in some cases. Because in many cases the investment opportunity will be in a product not currently produced in the country, host-country feasibility studies may need the assistance of industry-and product-specific know-how probably not available in the country.

One suggestion is formation of a nonprofit organization along the lines of the IESC, which could provide the specific expertise for feasibility studies. Or it could be a modification of a number of existing nonprofits whose work may sometimes involve feasibility studies, such as VITA. There are many consulting firms prepared to do feasibility studies at various levels of expertise and thoroughness, engineering consulting firms among them. But their job is not to eliminate the need for their services by means of technical aid to host-country personnel, although training, experience, and turnover of nationals contribute toward this end. Nor is it to help an LDC decide on specific feasibility studies. Perhaps VITA's new for-profit subsidiary, which seeks out investment opportunities in LDCs mainly for smaller firms in the U.S. illustrates one link in the chain. VITA's information service on available technologies illustrates another, and VITA/IESC technical assistance is a third. The new U.S.–ASEAN Center for Technology Exchange seeks to combine all three.

Indonesia

Given the stagnation of U.S. investment in Indonesia (natural resources extraction excepted), it must follow that investment missions and investment promotion efforts have not been successful. Although suggestions might be made to improve outreach to potential U.S. investors (and in fact the Indonesian government has contracted with three consulting firms to locate U.S. joint venture partners for Indonesian firms), this does not strike us as the priority or productive approach.

It is our opinion that these concerns are largely for the future; that at present they do not constitute a significant constraint on foreign invesment and therefore do not deserve high priority; and that investment in developing such capabilities now is premature, and would not offer high payoffs. First it is necessary to improve the investment climate, to reduce business uncertainty, and to lower the costs of entry and of doing business in Indonesia. What needs to be done to induce more U.S. investment is to persuade investors that the costs of the business environment in Indonesia are acceptable, even attractive. This is essential, as the possibilities for further investment solely for the Indonesian domestic market are limited at this time, and investment for export is highly sensitive to costs. This is not a promotion effort. What matters is not what investment promotion officers say, or what investment missions and government officials say. Those who must be persuaded are the U.S. firms already in Indonesia, the American Chamber of Commerce in Indonesia, and OPIC. They surely will be queried by any prospective investor, and their advice is critical. It is clear that the message conveyed by U.S. businessmen to their compatriots considering investment in Indonesia is discouraging—more a warning than a welcome. Japanese businessmen may be less outspoken, but some firms have withdrawn, business organizations have expressed concern, and investors have held back.

New investments must be approved by BKPM. But the information for making sound decisions, as well as the information for making feasibility studies by whatever party, is lacking. What is needed is an improved system for collection of current information on areas of business open to potential investors, current production in relation to demand, existing productive capacities, and market potential and location (Suriadjaya 1983, 18ff.).

What U.S. and Japanese organizations can do has already been discussed in terms of their contributions to absorptive capacity, and to managerial development and technology-specific training. This training is aimed at reducing labor costs (which are not low despite low wages), and at reducing the costs and uncertainties of government regulation (which remain high despite government reforms to simplify and expedite decisions). Once the costs of entry and doing business in Indonesia are reduced, more foreign investment will flow in.

Thailand

Investment promotion efforts have suffered from a lack of focus. It is clear that most investment feasibility reports are inadequate bait for the prospective investor; they have been prepared on a shoestring and do not provide enough product information. There is some disagreement as to the way to go. Adequate persuasive feasibility studies are expensive; the most effective feasibility studies are those conducted by the prospective investor. Many of the Thais capable of conducting such studies are reported working for foreign firms, many of them overseas. OPIC insists that such studies be done by the investor, and helps finance them. One suggestion is that there be "opportunity" studies, going into greater depth than existing feasibility studies.

BOI decisions appear to be made ad hoc on a case-by-case basis in response to proposals by business. The decisions do not appear to reflect either a budgetary decision on how much to spend (that is, how much revenue to forego), how much to allow prices to rise, how to allocate resources systematically across industries and products, or any benefit-cost estimating procedures. Nevertheless BOI should have some impact on the investment growth rate and on differential growth rates across industries and products. If its powers are to be effective, it must have a much stronger independent capability for feasibility studies than it now has, and a clearer sense of specific industrial development priorities. Feasibility studies are available to Thailand from various foreign sources free of charge. But these studies have a hidden agenda that is not Thailand's. Even if they had an appropriate agenda, they do not add up to a set of priorities or an investment incentive plan that could be presented to Thailand by any international donor. What donors can do is provide technical assistance in making feasibility studies.

The BOI needs to be more selective in the type of industry it seeks, perhaps in the type it promotes, that is, subsidizes. This appears to be the way BOI is moving. Whereas some previous investment missions to the United States were quite diffuse in their targets, their latest mission focused on electronics and targeted several cities that are centers of electronics manufacturing. Criticisms have been heard of the effectiveness of the U.S. Department of Commerce and OPIC in supporting investment missions to the United States (and to Thailand). Such criticism is the inevitable result of the diffuse nature of the missions themselves. Overseas missions also will be devoting much more attention to potential large-scale buyers of Thai products who might establish longer-term contractual relations with Thai producers.

Given the surplus of well-trained college graduates, including those in engineering and applied sciences, Thailand is in a position to attract and develop industries with greater requirements for technology. Thailand can support infrastructure that imposes new and more demanding requirements for absorptive capacity. A number of priority areas have been suggested (Yuthavong et al. 1985): biotechnology; development and conservation of land and water resources; electronics and information technology; and metallurgy and materials science. Further narrowing of priorities will identify specific needs in terms of materials standards, testing laboratories, and training and technical assistance needs, to which U.S. and Japanese can contribute.

Protecting Industrial Property Rights

One of the loudest complaints of U.S. firms is the lack of protection of industrial property rights—patents, trademarks, and licensing agreements. Indonesia has no patent law; Thailand does. But implementation is so slow and unsure, and compensation so inadequate, that there is little deterrence for violators, and no protection for the foreign firm. Trademark protection is also ineffective. Copying of products and trademarks is widespread. The question here is, how does this affect technology transfer? What can Japan and the United States, as distinguished from the countries in which violations occur, do about it? And is it in the interest of either Thailand or Indonesia to protect industrial property effectively?

Some foreign investment is deterred by lack of trademark protection. Firms are reluctant to invest in productive facilities and local market development if they risk appropriation of the benefits by local firms that pirate their trademarks. This is a problem in a wide range of consumer goods, computer software, and pharmaceuticals. Piracy occurs more in Thailand than in Indonesia. Although many products easily copied may not incorporate much potential for technology transfer, foreign investment does provide management training and technology transfer via training and turnover. There is also the risk that the product sold under a foreign firm's name is poorly made, for example, a dangerous or ineffective drug, or hazardous electric products. The firm may suffer from legal damages, and may lose valuable goodwill. Pharmaceutical firms in particular hesitate to introduce some new products in LDCs, including Thailand and Indonesia, for just these reasons.

When patents are involved, or substantial proprietary know-how, the lack of prompt and effective protection renders the foreign firm reluctant to train local employees and transfer technology, as the firm is only subsidizing potential competitors. Firms with such proprietary technology will be deterred from investing at all, or will be averse to consider joint ventures, minority ownership, or limitations on the employment of expatriates. They are also deterred from licensing their products or processes to domestic firms.

Thailand has no restrictions on licensing agreements, either in respect to the terms of the agreements or in terms of the technology imported. Whether or not this is in its best interests is beyond our scope (United Nations 1984, 232–33). Indonesia on the other hand has what are considered unduly restrictive limitations. Typically agreements are limited to five years, and payments to 2 percent of net sales. The five-year limit, renewal of which is by no means assured, effectively rules out transfer of technology with a long remaining economic life, even were there no concern about the lack of protection for patents or effective protection for trademarks. This lack provides an incentive for foreign firms to limit their transfer of technology and know-how so that they will not be faced with competition from their own former licensee once the agreement is terminated.

What can the United States or Japan do unilaterally or jointly? Unless the violator has a presence in Japan or the United States, there is little these countries can do, beyond protesting to the government in whose country the foreign business has been harmed.

If Thailand or Indonesia wish to protect industrial property, then the United States and Japan can provide consultants and technical assistance in setting up a patent office and implementing a process of patent review and protection. The plain fact is that it is not in the interest of Indonesia or Thailand at this stage of development to afford adequate protection for industrial property; they have more to gain from ignoring these property rights. This does mean that there is no way in which the latest technology, or any technology with a long remaining economic life in the advanced countries, will be transferred. The lack of a patent law in Indonesia speaks for itself. Products in violation of trademarks or patent law apparently are not being exported to Japan or the United States except in the suitcases of foreign travellers, nor have the violators seemed to have developed other export markets of any significance. They may both feel they have something to gain from the enterprise of local copiers.

Indonesia is not receiving foreign investments that require protection of proprietary knowledge. The exception is a few public firms; these receive technology mainly through licensing, not by foreign direct investment. Thailand is in a different position. It has the capability of attracting industries and developing industries for which proprietary technology is important. But in order to do so it must be able to demonstrate in advance to potential foreign investors that it is willing, as well as able, to provide prompt and effective protection. It takes years for adequate protection of industrial property rights to have its intended effects. The legal and administrative system has to be in place and adequately tested through the judicial system before it acquires credibility. It is the process of enforcement that matters, not merely the letter of the law.

The issue of protection of industrial property rights does not arise for those technologies transferred by the engineering consulting firms that design and build a plant and train local workers to operate it and maintain it. This is always hand-me-down technology. The latest technology is likely to be patented or inextricably linked with specific firm know-how. Such technology will only be transferred if the firm's property rights can be adequately protected, whether through licensing or through direct investment and production in the host country. Open technology embodied in turnkey plants and complementary training could be the best technology for a particular LDC, or it could be acceptable technology if the product is only to be supplied to a protected domestic market.

To export products, it may be necessary to have access to proprietary technology, and to have access to (or capability for) continuing process and/or product modification. This implies continuing technical assistance via licensing or direct investment by the possessor of the patent, trademark, or special know-how. If in fact industrial property rights lack assured protection in an LDC, then direct foreign investment may be the only option until the LDC has the capability to reverse engineer and develop independently. No country, including the United States, does this across the board.

Increasing Support for Nonprofit Organizations

The unique characteristic of nonprofit voluntary organizations is that they are autonomous. They are not subject to outside political influence, or to national considerations. Interpretation and implementation of their goals is achieved through internal discussion. The fact that their origin is in the United States, or that their headquarters or even their main source of funds are in the United States, is irrelevant. Some of these organizations receive financial support not only from AID, but from the foreign aid organizations of other nations, from international organizations, and from the governments of the host countries.

Some of these organizations could use more financial support, and would use it more effectively for economic development than the unavoidably political foreign aid organizations of the United States or Japan. AID is already contributing to a number of them. What must be avoided is dependence by a nonprofit on any one donor for financial resources—otherwise its independence may be compromised. The longer the time period over which outside funds are assured, the smaller the risk of dependence.

As indicated before, there are three kinds of nonprofits whose activities contribute to human resource development relevant for technology transfer. The first kind contributes primarily to absorptive capacity. The Ford and Rockefeller Foundations, and the Agricultural Development Council, are prime examples. They provide high-level education-training, consultants, visiting professors, and research support. Some of the larger foundations in fact contribute to other nonprofits in the same manner as we suggest that the Japanese and U.S. development agencies might contribute to them. Many foundations, however, may be unwilling to expand the scope of their activities using outside funding sources not completely under their control. Others, less well-endowed, would be pleased to receive additional funds.

The second group of nonprofits, exemplified by IESC and VITA, are engaged in technology-specific training and technical assistance and, in the case of VITA, in technology information services matching client needs with available technologies. The primary need in their case appears to be not additional funding, but dissemination of information about their availability and the services they offer. IESC has a list of some 10,000 retired executives available worldwide, and would be able to meet far more requests than it gets in Indonesia and Thailand. VITA has a computerized file of technologies which is essentially a public good. More clients could make use of it without reducing its availability to other clients. Some other services, such as technical assistance and development of appropriate technology, are more limited and more expensive.

Interviewees in the countries studied claim that the cost (although heavily subsidized) is the main factor in underutilization of IESC. This might be a constraint if the availability of the services were more widely known, but at present it is not. IESC philosophy is that the client should pay according to ability, if the client is to be motivated to make good use of the consultant's services. One might experiment with reduced payments (that is, increased subsidy), but lack of information appears to be the immediate cause of IESC's underutilization.

Technology, of course, is more international than advisers and consultants. In fact a good deal of VITA's technology inventory is not from the United States, and some has been developed with its financing in LDCs by citizens of LDCs. What is needed first and most is a concerted effort to identify sources of technology and related technical assistance available free or well below cost to LDCs and disseminate this information to potential users.

In sum, there is an excess supply of technology information and of technical assistance available through nonprofits. Government and business organizations should work with them to spread the knowledge of their availability more widely.

As to selected nonprofit organizations whose main contribution is to community development and the like (primarily in rural areas), their mission falls outside the scope of technology transfer. However, modification of values and attitudes, provision of information, example, and training of leadership, all of which must be conducive to change, encourage a heightened awareness of and openness to new technology. In fact nonprofits' activities in rural-agricultural-community development are relevant to the need to generate large numbers of new small firms in rural areas, preparing the ground for transfer and diffusion of technology, and to greatly expand employment and earning opportunities there. They help lay the groundwork for the activities of the bank loan officer. They do help identify investment opportunities, and they help to identify and train local leaders, some of whom might be successful entrepreneurs. Thus nonprofits can help the bank loan officer in generating demand for credit. The loan officer in turn is in a position to increase the credit available for local community development efforts insofar as they can meet some test of ability to repay. The private nonprofits supply some technical assistance, and might contribute to, and help seek, technical assistance resources tapped through bank loan officers. Such prospective cooperative efforts might in turn help wean some nonprofits away from excessive concentration on relief to those activities that would eventually end the need for further relief.

Thus, training and technical assistance by foreign firms and governments to suppliers of credit, and financial assistance for private nonprofits, could have a synergistic effect.

Alwyn Young (Young 1984, 12ff.) argues that private nonprofit organizations can help bridge the gap between the poor, the informal sector, government programs available to the formal sector, and assistance from private business. Nonprofits can be assisted to prepare employment and small business programs by international aid donors to cover initial operating costs, and by technical assistance. Nonprofits have better access to potential entrepreneurs than government agencies and can recruit better personnel. In particular they can reach small business, whereas government programs are almost exclusively concerned with industry. Young's assumption, shared by Oshima (Oshima 1984), is that industry has very limited possibilities for growth in employment, and that generating employment and increasing incomes of the mass of the population must be accomplished via the stimulation of small business, not industry.

The possibilities for linkage between the activities sponsored by private nonprofits are greatest perhaps in the field of agriculture. In Indonesia many farmers are beyond the reach of extension workers. Government programs for agricultural and rural community development are only available to villages that are well-organized and have the information required to take advantage of existing programs. PVOs can promote village organization and facilitate access to sources of new technology and of the credit and technical assistance required to make good use of it. PVOs, in helping introduce new crops and new local industries, prepare the ground for local entrepreneurs who can establish small businesses and modernize existing ones.

What can the U.S. and Japanese governments do? First, they can provide selective financial support to permit the expansion of activities by private nonprofits that contribute to absorptive capacity, to specific training and technology transfer, and to community development. Second, they should promote linkage (via information and possibly financial incentives) between nonprofits engaged in community development and local and foreign organizations whose focus is on training entrepreneurs, financing and providing technical assistance to small business, and providing suitable technology. Third, they may have a contribution to make as collectors and purveyors of information on available technologies and technical assistance from diverse sources.

Japan has been rich enough, long enough, to develop its own private nonprofit organizations contributing to human resource development and technology transfer to LDCs. In the United States, initiative has come from many sources, but the tax system has been a major inducement toward endowment of nonprofits. Japan may take different routes, but should now develop institutions not bound by either political or profit considerations.

Recommendations to U.S. and Japanese Firms and Business Organizations

Most technological transfer is done by business, with technology-specific training of nationals being an essential and integral part of the transfer. The main contribution of governments and nonprofits to human resource development facilitates technology transfer by business, by improving government administration, the educational system, and the functioning of economic and social infrastructure of transportation, communication, utilities, and public health services. Foreign investment (and investment in higher-technology production) when the preconditions for success have not been met must be at a high cost to the recipient country—in terms of subsidies to foreign investors, and in terms of tariff and other forms of protection that raise prices to consumers (and business).

In 1973, the U.S. National Academy of Sciences published a report on technical cooperation with developing countries (National Academy of Sciences 1973, 43–44). Its recommendations to international business firms on cooperation in human resource development bear repeating. Although directed at MNEs, the recommendations imply action on the part of the host nation's institutions if they are to be implemented.

Train more technical and managerial personnel than the firm requires

Hire qualified graduates from developing country universities

Subcontract as feasible

Use developing country technical and scientific services to the extent feasible

Extend relations with developing country universities and technical schools

Consider local contracts for research projects

Use local faculty as consultants

Make corporate personnel available for university teaching and other services

Aid local professional societies

Assist local supplier industries in meeting product standards

Some of these recommendations refer to absorptive capacity, and others to technology-specific training and technical assistance. All are being implemented by some firms to some extent. Business should consider whether firms could do more. What is reasonable for a firm that is profit-oriented to do depends upon the local environment to a considerable extent. But it also depends on taking a longer view rather than fixation on the next quarterly or annual report. This is a problem for U.S. firms, not Japanese. Japanese short-sightedness is of a different kind: slowness to rely on initially inadequate local institutions and firms.

Other recommendations also bear indirectly on technology transfer and its human resource component.

Conduct research, development, and experimentation for labor-intensive, scaled-down equipment

Reduce the number and variety of product options

Ease performance standards in noncritical (not affecting safety or quality) areas if costs can be reduced by so doing

Modify processing techniques to use more local materials

Eliminate expensive packaging materials

The last group of NAS recommendations would accomplish several things. They would help reduce the need for technology transfer and related training associated with local production by foreign firms; expand the prospects for technology transfer by developing "appropriate" technology; and increase the host-country benefits from new products and processes by increasing the domestic value added, thus reducing import requirements.

Our own additional but related recommendations include the following.

1. Increase the amount of direct investment.
2. Increase direct investment in exportable products, whose market is larger than the limits set by domestic markets.
3. Change composition of direct investment toward higher-technology investment in terms of products, and in terms of productive processes.
4. Stress investments with many linkages, so as to induce further investment in supplier industries and in industries using the product as a major input.
5. Increase licensing and associated technical assistance for domestic firms.
6. Make better use of expatriates as agents for implementing technology

transfer. U.S. firms may use too few expatriates, and certainly their assignments in most cases are much too short. Career incentives should be improved so that more expatriates are prepared to spend longer assignments in the host country. Japanese firms tend to use too many expatriates, and the rate at which they are replaced by local managerial and professional employees is much too slow from the viewpoint of technology transfer to the host country. (There is an optimum rate of replacement, because expatriates are important for training domestic workers, and too-rapid replacement hampers training, productivity, and output rather than increasing technology transfer.)

7. Increase turnover of trained and experienced workers to the domestic economy (again, there may be some optimum; too high a rate implies high training and therefore production costs, thus discouraging investment, exports, output, and training). This is accepted by U.S. firms, but runs counter to the management philosophy of Japanese firms.

In order to bring about desirable changes in numbers 1-6, improvement of incentives and reduction of constraints is needed. In every case, this may involve expansion of appropriate absorptive capacity through general education, increasing the supply and quality of workers, reducing training needs and costs, and thereby lowering production costs. Higher-technology products and processes would become economical. Export possibilities would open up in some cases. Higher absorptive capacity allows more investment; creates opportunities for turnover to the domestic economy of trained and experienced workers; facilitates replacement of expatriates; and opens up economic possibilities for higher-technology investment and for licensing of higher-technology products and processes to domestic firms.

As to number 7, turnover, it is not entirely in the control of individual firms. It is hardly ever in their interest to increase turnover, whatever the gains for the economy. Growth of investment and production will raise turnover above the very low rates that reflect lack of alternative opportunities; increasing absorptive capacity and technology-specific training will reduce very high rates resulting from "raiding" in a labor market critically short of specific skills and experience.

Legal protection for industrial and intellectual property, where inadequate, is a constraint on technology transfer, especially in numbers 3 and 5. This, however, is the responsibility of the host country.

Underlying most of these initiatives are the level and type of investment by U.S. and Japanese firms, and in the domestic economy. Continuing foreign investment increases training activities as well as demand for domestic goods and services. This results in growth of absorptive capacity and increased technology transfer. Domestic investment and growth are essential to provide opportunities for experienced managers and professionals to implement the "foreign" technology that they have mastered in domestic organizations.

Now is the time to stress foreign investment as the priority in accelerating technology transfer for Thailand; this time is still in the future for Indonesia, with the exception of oil, gas, and minerals. In Indonesia we are recommending more training on an industry-wide basis by foreign firms because of the inadequacies of the Indonesian educational and training system, and in particular the lack of well-qualified teachers in the nontraditional skills needed. In Thailand it seems appropriate to stress the contribution firms can make to local educational and training institutions. This is because the latter are much better prepared to train for the skills needed by business and because of the greater reliance on hiring experienced workers, ás far as U.S. firms are concerned. Many U.S. firms have no (or minimal) training programs on that account. This does not preclude some on-the-job training and experience by firms for nonemployees. But this should be regarded as a complement to local educational programs, not as a substitute for them.

To the extent that foreign investment (and U.S. investment in particular) is deterred by local ownership requirements, U.S. and Japanese firms already in Indonesia and Thailand can help develop a broader local capital market, with greater protection against manipulation and instability. Both Indonesia and Thailand have incipient capital markets, but only 20 companies are listed in the former, and about 70 in the latter. Increasing the number of firms with listed securities offers attractive opportunities to local investors but with domestic ownership shares sufficiently dispersed so that management control is not threatened.

In both countries, but particularly in Indonesia, U.S. firms are represented predominantly by large multinationals, limiting the scope and diversity of U.S. investment. This is not an issue with Japanese firms because of the role played by the *sogoshosha* and JETRO. Steps should be taken to facilitate investment and technology transfer by smaller U.S. firms. From the viewpoint of the host countries, smaller firms have a number of advantages over large multinationals. They are more willing to accept a minority ownership position; they are more willing to license technology to local firms rather than insist on direct investment; and they are typically in a weaker bargaining position.

What smaller firms lack is information, support services and facilities that larger firms provide for themselves, or can more easily contract for. The U.S. organizations that have the capability for correcting these deficiencies, thus making LDCs more accessible to smaller American firms, have not yet been developed. Until the passage of the Export Trading Act of 1982, they could not perform that role. The organizations best equipped to fill the gap in information are the large U.S. banks. The economy-wide range of their leading activities requires them to be knowledgeable beyond the needs and experience of any particular industry. The banks can develop sources of information that can be used for evaluating new prospects and new markets. Beyond supplying information, they are obvious sources of credit. They are less well equipped to provide other services that may be needed by smaller firms: training, marketing know-how in LDCs, or technical assistance. The spread of counterpurchase policies among many LDCs, including Indonesia, presents an opportunity for banks to extend the range of their activities and know-how into international trading of LDC products. Essentially, firms are required to export additional amounts of some LDC product (typically unrelated to their own line of activity) in order to obtain permission for imports of equipment and supplies of the same amount. Some banks, with their extensive information resources and international branches, have offered to handle this business for a fee.

Another type of organization that can compensate for some of the short-comings of smaller firms is a large U.S. retailer, such as Sears, Roebuck. Sears can provide a market for exports of consumer goods to the United States, and to other countries where Sears has outlets. It has long experience in developing contractors in LDCs, including provision of credit and technical assistance. This type of organization is more limited in scope than the bank, because it is only interested in consumer goods, and in markets it can reach through its stores, which are primarily export markets to the United States. But it can provide assured markets, which banks cannot do. It lacks the local knowledge of the bank, but is in a better position to gain access to it than the individual small firm. In some countries, Sears has established a network of stores, and has assisted local manufacturers in supplying products for Sears' domestic sales (typically encouraged by tariffs on imported goods), and has sometimes expanded into exports. However, this activity is ruled out by Indonesian law, which reserves sales to local firms. Although a joint venture arrangement would be feasible, and the Sears trademark could be used, neither Sears nor other large U.S. retailers seem to have developed either local producers, or induced U.S. firms in Indonesia or Thailand to become suppliers.

In addition to full-service intermediaries somewhat analogous to the Japanese trading companies, there is a role for so-called technology brokers who match seekers and suppliers of technology. Although such brokers exist, their contribution in Indonesia and Thailand is barely beginning. Baranson (Baranson 1981, 143ff.) has suggested creation of a technology transfer service corporation to bridge the gap between would-be U.S. supplier firms and LDC firms seeking technology packages. To a very limited extent, the U.S. nonprofits already mentioned are filling the information needs. For financing, however, there is need for the full-service intermediaries allowed under the Export Trading Act.

Recommendations to the U.S. and Japanese Governments

The Asia–Pacific Council of American Chambers of Commerce (APCAC) at their 1984 Singapore meeting adopted a number of resolutions, some of which consist of recommendations to the U.S. government which, if adopted, would encourage technology transfer (*Thai–American Business* 1984, 12–18). With regard to industrial and intellectual property rights, APCAC recommends legislation making trademark counterfeiting a criminal offense; negotiation of international codes to discourage international traffic in counterfeit goods; assisting individual LDCs to strengthen intellectual property protection; and holding host countries accountable for misappropriation of or inadequate protection of intellectual property rights.

On domestic content, APCAC opposes domestic content legislation in the United States or anywhere else.

On export financing, APCAC recommends that the Export-Import Bank's lending authority should realistically reflect market demand and should be competitive in terms of interest rates, contract coverage, and fees; that the bank should develop a financing and guarantee program for leases, and a program of medium-term credit. On private banking services, APCAC urges the U.S. government to seek "national treatment" for U.S. banks in countries where the banks' opportunities are restricted. It also recommends expansion of the Trade Development Program (TDP) and increased funding to provide U.S. consulting, engineering, and construction companies access to world-wide markets.

Finally, APCAC calls for Congress to amend the Foreign Corrupt Practices Act along lines that would restrict its application to bribes and exempt normal "grease" payments.

Although some of these recommendations refer to trade rather than to investment, they are applicable to investment as well. Trade, although less than investment, embodies some technology transfer and often involves some training and technical assistance.

Lawrence Krause, discussing U.S. policy toward ASEAN, finds that a number of U.S. laws and administrative procedures have caused a deterioration in nonprice elements of trade competition (and, by inference, also in the conditions for direct investment). He mentions among others the Foreign Corrupt Practices Act; the taxation of U.S. citizens residing abroad; introduction of human rights criteria in Export-Import Bank loans; introduction of environmental impact criteria; rules relating to procurement under AID programs; and the extraterritorial reach of U.S. antitrust laws (Krause 1982, 53, 61).

In listening to the catalog of complaints by U.S. businessmen abroad with regard to U.S. government policy, one needs to determine to what extent they are a reaction to pesky irritants—"hassle" factors —and to what extent they pinpoint significant deterrents to investment and technology transfer by U.S. firms. The loudest and most widespread complaints encountered in Indonesia and Thailand refer to the Foreign Corrupt Practices Act. The act is an acute embarrassment to U.S. businessmen both because of statements required of local business contacts, and because of local resentment of the extraterritorial reach of U.S. law. It involves some reporting costs, generates some uncertainties, and places U.S. firms at some disadvantage in competing with firms not subject to the act. The act is not believed to inhibit greatly the performance of U.S. firms, but there is no way of being sure about this. What is certain is that the act does not distinguish between "grease" or "speed" payments (which are an accepted way of doing business, what U.S. firms might call "user charges") and bribes and practices considered corrupt in the host country.

The second most prevalent complaint refers to lack of protection of industrial and intellectual property rights. It is louder among U.S. than Japanese businessmen, but both countries are affected. Although the U.S. and Japanese governments might exercise some influence, ultimately it is the Indonesian and Thai governments that must provide protection.

With regard to AID, the level of funding (particularly in Thailand) is low compared to other donors (and specifically to Japan), and low relative to previous levels. Much of AID's activity is aimed at developing absorptive capacity. AID could well consider more direct assistance to training activities of U.S. business firms in Indonesia, until such time as Indonesian educational and training institutions are capable of taking over. Such assistance has precedents in federal government manpower training policies. JICA should increase its contributions toward expanding absorptive capacity in Indonesia, primarily through the educational system and administrative structure. In Thailand, JICA could contribute selectively toward developing research capability and technological infrastructure. Not enough of JICA's activities are committed to these long-term developmental objectives. Other U.S. public organizations, the embassy and the Commerce Department in particular, could provide greater support for U.S. investors and work more actively in their behalf. Alternatively, the activities of OPIC should be expanded.

If most of these recommendations apply to the U.S. government only, it is because the Japanese government (its embassies, JICA, and JETRO in particular) is working very closely with Japanese firms and is not imposing the constraints or embarrassments on Japanese firms that U.S. legislation has imposed on U.S. firms.

The Scope for Japanese–U.S. Cooperation

Private Nonprofits

It should be possible for some nonprofits from the United States and Japan to cooperate. Because this is an area in which Japan lags, the scope depends upon the development and expansion of nonprofit organizations in Japan. Obviously this is not possible across the board. Some nonprofits are peculiarly American; others have internal goals for which there is no Japanese counterpart (for example, religious organizations). But most of the private nonprofits engaged in activities relevant to technology transfer are American primarily in historical origin and headquarters location; they employ foreign nationals, receive funds from foreign governments, in some cases contribute to other nonprofits head-quartered in countries other than the United States. They are free of political influence, U.S. or other, in choosing their objectives or in designing and imple-menting programs. There is scope for cooperation, particularly in expanding absorptive capacity, in institution building in applied research, and in developing a technological infrastructure. This can take the form of supplementing the resources available to existing U.S. nonprofits and/or complementing and divi-sion of labor with U.S. organizations.

It must be kept in mind that U.S. nonprofits are numerous and diverse, and independent of one another except for some cross-financing. There is no associ-ation with which to interact (although AID has a Private Volunteer Organiza-tions Office). Cooperation must start on a one-to-one basis. In fact, Toyota Foundation has been in contact with the Ford Foundation in Jakarta.

Business Firms

There is little prospect of cooperation between U.S. and Japanese firms in activities enhancing technology transfer. Neither group is interested in technol-ogy transfer per se. Their industry distribution is quite different, and to the extent that it overlaps, it is highly competitive. U.S. businessmen feel that they have nothing to gain and much to lose via joint ventures with Japanese firms in Indonesia or Thailand; they do not feel that they play by the same rules. They also feel insecure in their arms-length relation with their own government, contrasted to the symbiotic relation between Japanese business and Japanese government agencies. Essentially, U.S. businesses in these countries lack the capacity to act together except in the common defense, whereas Japanese business is much more cohesive with regard to the outside world (notwithstanding vigorous competi-tion between enterprise groups).

What basis is there for cooperation between Japanese and U.S. business? Foreign business communities have a common interest in increasing the absorp-tive capacity of the domestic economy by improving the quality of education, and by developing the infrastructure of information, communication, legal systems, transportation, utilities, and standards. There has been occasional cooperation in lobbying for regulations considered beneficial, but generally they have united to defeat harmful regulations. They can contribute (with personnel, cash, and equipment) to improving the quality and quantity of education and training, as some have been doing individually. Or they can contribute through industry associations, and could consider cooperation with government agencies and nonprofits engaged in the same or similar activities. At the the industry or firm level, the main scope is for training in industry-specific skills. This can be done (and has been done) by particular firms with the support of others, or by industry associations for their members. In either case it has been done with cooperation of appropriate government institutions.

It is easy to recommend closer cooperation between U.S. business and government. The remote if correct relation between U.S. business and the U.S. government is reciprocal; business feels that their government is not well informed nor very interested in their problems—and as far as development agencies are concerned, they have functioned until recently in isolation from the business community. Government agencies on the other hand cannot expect the cooperation of business that is taken for granted by the Japanese. Government feels that it must be impartial among U.S. business interests and firms, and the requirement of impartiality is interpreted as noninvolvement.

JICA serves Japanese business interests in a way that AID does not and never will, and the interests served are only indirectly related to technology transfer. AID is more pluralistic, is far less focused in its activities than JICA. Its interface with U.S. business in LDCs is only recent (or on again, in a long history of on-and-off-again relations) and still limited. AID has nevertheless changed its focus toward a greater stress on private enterprise and private investment as agents for development and technology transfer. Before AID can become truly effective, time must pass in order to confirm the longer-term nature of AID's commitment to work with and through business. Also, experience and trust must be built, a process handicapped by high rates of turnover both of U.S. businessmen and of AID personnel as well.

Certainly more interaction with U.S. firms in LDCs, and with prospective investors, is a logical step. There is a clear division of labor between AID and OPIC—the latter is concerned up to the point an investment is made, and AID may largely pick up at that point. OPIC is concerned only with specific feasibility studies, projects, and complementary training; AID is concerned with larger issues of absorptive capacity.

Government Agencies

Because of the large differences in program activities between JICA and AID, it may appear that without major alteration in the programs or policies of one or the other, the opportunities for cooperation are absent. This, however, is not the case. There are projects, industries, and sectors where they may complement each other. The possibilities for cooperation are much more realistic in terms of increasing the absorptive capacity of an Indonesia, or in terms of strengthening the technological infrastructure of a Thailand, than in terms of investment in specific productive facilities and in the associated specific training—an activity in which competition is likely to prevail over cooperation.

The main areas offering opportunities for cooperative activities are agriculture and natural resources. The fisheries industry is of particular interest to Japan. Both Thailand and Indonesia need assistance in (1) establishing data bases via oceanographic surveys; (2) creating the facilities for regular monitoring of fisheries resources, conditions, and activities; and (3) in establishing or strengthening applied research capabilities to improve fisheries, and fish and seafood processing. This includes aquaculture, which provides opportunities for controlled increases in yield.

The forestry industry is also of particular interest to Japan, the main consumer of logs and wood products. Here some improvement in information on available stocks is needed, but the main requirements are for research on conditions for sustained yield, methods of improving yields, and the impacts of deforestation on the environment. In Indonesia, main concerns are the effects of cutting practices on soil erosion, the depletion of humus and soluble minerals, and the siltation of streams and irrigation systems. In Thailand the problem is more one of long-term replenishment of forest resources; the impact of cutting mangrove in tidal areas on shrimp and other fisheries; and minimizing the agriculutural impacts of the deforestation that has already occurred. In both countries, especially in Thailand, there is the need to minimize further deforestation at the local village level by providing viable alternative sources of local fuel and building materials.

Agriculture, which employs the majority of the population in both countries, is clearly the top priority for development assistance. Many current and potential crops are of interest to both U.S. and Japanese firms which, through contracts with farmers, guarantee them a market and also provide a variety of technical and other help and information on quality control. But they reach only a small proportion of farmers in Thailand, and a tiny proportion in Indonesia. These firms cannot be depended on to provide similar help for crops that they are not interested in exporting, or for that share of export crops that is consumed domestically. Again the development and strengthening of applied research facilities is essential. This involves the education and training of a wide variety of specialists on plants, pests, soils, and water management. In some cases the numbers needed do not justify developing a quality domestic educational capability; in other cases they do.

A number of international research centers already provides useful information of value to Thailand and Indonesia. However, internationally generated information and options must be supplemented by location-specific research, taking into consideration local soil and climate conditions, local diseases, and pests that require location-specific adaptations. It is necessary to have rapid response capabilities because of the quick spread of diseases and pests under tropical conditions, the development of local resistance to specific insecticides, and the peculiar characteristics of local plant varieties. Irrigation farming, particularly with the new rice hybrids, requires a synchronization of planting and harvesting that aggravates the problems of pest control and imposes large management requirements on agricultural production: management of water, peak load labor, storage, and transportation requirements. Although research facilities may not include management responsibilities, their advice implies management practices that in turn require an increased number of trained and effective intermediaries.

Assistance in developing needed research capability involves not just the provision of facilities and equipment and the training of research specialists. It also involves improving local educational capabilities to train "middlemen", individuals who can communicate (and translate if necessary) and advise on the implementation of research findings, and who also can communicate to the researchers the needs and priorities from the viewpoint of the producers.

The one instance of Japanese–U.S. cooperation is precisely development of research and related support facilities at Khon Kaen University in Northeast Thailand. This, however, has become more a parallel effort with frequent communication than a truly cooperative effort. The division of labor follows the differences in emphasis of AID and JICA, with the United States contributing more to the absorptive capacity and high-level training, and the Japanese contributing equipment and technology-specific training. The United States works in close cooperation with the university, and the Japanese choose to work with the Ministry of Agriculture's Department of Land Development.

It is easier to specify in what areas U.S.–Japanese government cooperation is unlikely than those in which it is feasible. In projects that are likely to result in exports to the host countries, competition for trade may rule out cooperation. Research, information, organization, and planning assistance typically do not. And host countries are less likely to have reservations about cooperation in these areas. Research on volcanology and seismology is a Japanese strength; warning systems, including tsunami warnings, involve communications equipment. Bangkok's traffic and drainage problems can be attacked by cooperative efforts up to a point. At the design stage, where imported equipment becomes involved, cooperation becomes difficult, unless Thailand is in a position to make independent technical decisions.

One area for potential cooperation is education. U.S. government contributions have concentrated on more formal, higher education, whereas the Japanese efforts have focused on more vocationally oriented training. There is no reason why there could not be cooperation in the former; some of the latter may be too industry- or project-specific for cooperation.

Research, as indicated, is a second obvious area for cooperative effort. In addition to providing training, technical assistance, financial support, and equipment, the U.S. and Japanese governments could facilitate the flow of information on ongoing research.

Although conditions unique to particular countries call for indigenous adaptive research, there are common problems susceptible to common solutions. These could be addressed by the U.S. and Japanese governments. One type of appropriate technology is downsizing of productive facilities and processes to reflect small markets or small supplies of key inputs. Development of efficient labor-intensive technologies is a will o' the wisp, but products and processes that economize on scarce resources could be facilitated (management in the case of Indonesia, capital in most LDCs).

The role that both the United States and Japan can play is twofold. They can act as collectors and purveyors of information on available technologies other than those used in their home industries. And they can act as supporters of and possibly as participants in creation of new appropriate technology and in the creation of facilities and provision of resources with the capability of developing appropriate technology and disseminating it to Third World countries. Internal dissemination of technologies new to a country on the other hand must be accomplished through indigenous institutions and resources, although they may benefit from technical assistance from Japan and the United States.

A major need is information, and this is a third area in which the Japanese and U.S. governments could cooperate. Information flows could be improved on ongoing research, and on available technologies. Government agencies could assess the status of existing repositories of information on technologies and products, which include some government agencies, some specialized private firms, and some nonprofits. JICA and AID could not only improve the information network, in part by consolidating and maintaining a continuous inventory, but could also improve access by other means (by financial assistance, or complementing services). A similar inventory of technical assistance and consulting resources would complement the information network. Finally, they may contribute to the support of research on "appropriate" technologies so that each country need not try to duplicate the efforts of others.

There is a large number of separate organizations performing some role in technology transfer-related human resource development in Indonesia and in Thailand. The main donors do meet periodically and exchange information, although doubts have been expressed by some interviewees as to the adequacy of even this limited effort. Many other donors, whose combined contribution is not inconsequential, are going their separate ways. These represent not just the United States and Japan, but many other nations and multinational organizations. If anyone were to coordinate these efforts, it would have to be the host country, with its own large commitments to human resource development and technology transfer. AID has conducted surveys in Indonesia. JICA has conducted surveys for Japanese firms. Both countries should be able to cooperate in instituting a periodic survey of activities by foreign organizations (government agencies, firms, and nonprofits) not limited to those of their own nationality. Such surveys would be helpful to donors in designing their programs, and would provide the Thai and Indonesian governments with the information needed to coordinate the contributions of many agents with each other and with their own training and education programs and industrial policies.

Needs for information extend far beyond information on availability and sources of technology. Information about relevant aspects of the domestic economy is important for investors. In Indonesia, in particular, information is inadequate in quantity, quality, and timeliness. "[P]otential domestic investors [need] information on the areas of business open to them, the size of production compared with demand, existing production capacity, the location and potential of the market, the income distribution of the population, and so on. In Indonesia this type of information is still very scarce" (Suriadjaya 1983, 19). Most of the information must be collected and distributed by local government agencies, but the U.S. and Japanese governments can contribute, as they already are doing, by helping design systems for collecting information, by training personnel who can implement and manage data collection and analysis, and by providing technical assistance.

New investors first need information on markets for new investments. This is ultimately the responsibility of the investor, who has most at stake, but improved regularly updated statistics on consumption and on underlying determinants of demand would be very helpful here. Most of the contribution would come from agencies concerned with data collection and analysis, whether demographic, price or other market information. Private firms specializing in market analysis could also make a valuable contribution. Second, information is needed on prospective skill requirements of the economy. Most important are new skills whose local supply is highly inadequate and the training capability for which does not exist except in foreign firms. This is a matter for collaboration between foreign firms and/or their business organizations on one hand, and host government organizations concerned with labor training and education. For such "bottleneck" skills, timely information is needed on training capabilities and activities of foreign firms both in the host country and overseas, to determine what action host-country governments should take. This type of information should be obtained for all major new foreign investments, particularly those involving new products and processes.

Third, selected information is needed on costs, prices of inputs, wages, and productivity, all of which, in combination with market information, serve as the basis of estimating investment opportunities for the domestic market. Also, more critically, they influence the prospective competitiveness of local production in international markets.

Facilitating Technology Transfer: The Role of Indonesia and Thailand

The original objective of this book was to learn how U.S. and Japanese organizations (and inferentially those of other advanced nations) might improve and expand the process of technology transfer to Indonesia and Thailand (and inferentially to other LDCs). However, in the course of numerous interviews and discussions, it became apparent that some of the measures to encourage or facilitate technology transfer would have to be taken by the host countries. Further, it became clear that there were deterrents to technology transfer in the form of host-country regulations and administration as well as in conditions that were beyond the sphere of action and perhaps even of influence of U.S. and Japanese organizations.

It was realized from the start that much of what each country can do, and anything that the two advanced nations might wish to do jointly, would have to be with the consent and participation of the host countries. One conclusion of our research is that the main requirement is the removal or diminution of obstacles to technology transfer rather than the provision of assistance or incentives by Japan and/or the United States. The obstacles are predominantly found in the host countries themselves, and particularly in government policies that discourage technology transfer, although that is not their purpose. Thus, although our recommendations still refer to public and private U.S. and Japanese organizations, much of what they could contribute is assisting Thailand and Indonesia in removing or countering obstacles, including those of their own making. Much of the initiative must lie therefore with the Indonesian and Thai governments.

The recommendations we have to make in this regard are not new, and in the past have been made by host governments themselves, but have not been adequately implemented insofar as the same recommendations remain on the agenda.

In Thailand, no single deterrent to technology transfer appears to stand out. There is much grumbling by foreign business about regulations and their administration. The regulations are described as cumbersome and time-consuming, but are not unreasonable. Some uncertainty arises from jurisdictional conflicts among government agencies concerned with policies affecting foreign business. The system of side-payments is considered to be well institutionalized, its costs modest, and performance, reliable. Except in pharmaceuticals there are no complaints about unfair competition by government enterprises. Restrictions on participation of foreign firms in certain industries do reduce investment and technology transfer. The lack of effective protection for patents and trademarks is loudly decried. At this time it is not clear that much investment or technology transfer is being deterred thereby. But the structural changes that Thailand is planning would highlight industries for which protection of industrial and intellectual property is important, in particular inasmuch as they would depend on export markets.

In Indonesia, there is no question that the local business environment has deterred both U.S. and Japanese investors, and has been a major factor in decisions by U.S. and Japanese firms already located there to terminate their operations.

One important reason appears to be that prospective investors perceive the costs of business entry and business operation to be high. Some of the factors in the perceived high cost of business operations have already been mentioned. The inadequate educational background and supply of skills result in low productivity and high training costs. In some industries, inadequacies in infrastructure (transportation, communications, power, and other utilities) raise costs through delays and interruptions of production and delivery or cause firms to supply their own infrastructure at high cost. But another factor is mentioned and stressed at least as often: the costs and uncertainties associated with government regulation.

The process of negotiating all the approvals required to start business is complex, costly, and of highly uncertain duration and outcome. It is this uncertainty more than any substantial but calculable cost that deters new investors. Firms with long experience in Indonesia are better able to cope with the regulatory environment.

The main source of uncertainty (and the prohibitively high prospective costs into which it is converted) is an inadequately institutionalized system of side payments for public services and regulatory decisions. The prospective foreign investor, and to a lesser extent the local firm, domestic as well as foreign, is not always sure whom to pay, and may pay the wrong individual; it is unsure of how much to pay, and may be induced to pay too much; it is uncertain that services paid for will be delivered.

Top government officials speak out against these practices. Reforms have been announced more than once to reduce the delays and uncertainties that raise the costs of investment and business operation, and to reduce the number of "gatekeepers" and the associated opportunities for private gain of bureaucratic gatekeepers at the expense of business and government's own interests. But the management depth at middle and lower levels is lacking. This depth is essential to implement, monitor, and enforce decisions taken at the top.

It is widely believed that there is a lack of entrepreneurship in Indonesia, and particularly in Java. We do not share this belief, but suspect that entrepreneurial abilities have been channelled to bureaucratic roles and diverted from productive to predatory redistributive activities. Thus, reduction of abuses of the "gatekeeper" roles of government would contribute toward solving the problems of inadequate enterprise as well.

Two specific policies that may deter foreign investment, in particular investment with valuable proprietary technology, are limitation of foreign firms to minority ownership (a policy adversely affectng U.S. but not Japanese investment), and ceilings on the number of expatriate workers allowed and the rules on their rate of replacement. Both policies are far more sweeping in Indonesia than in Thailand, where they do not appear to have much effect on U.S. or Japanese investment. Indonesia has also begun to enforce its policies on expatriates far more rigorously.

It is not realistic to expect Indonesia to eliminate the requirement that all foreign firms (with few exceptions) have an Indonesian joint venture partner, nor that the Indonesian share of ownership must rise to at least 51 percent in 10 years. In fact one may expect a more effective enforcement of these laws in the future. But there is a price to pay, and the more valuable the technology sought, the higher the price. As indicated above, development of an effective equity market could reduce the deterrent effect for U.S. firms.

A second more subtle deterrent is a widespread attitude toward foreign business that regards it as a necessary but temporary expedient, to be kept on a leash, used, and dispensed with as soon as possible. It is reflected concretely in rules about employment and replacement of expatriates; limitations on foreign ownership; agitation against foreign trademarks; and restrictions on duration and payments under licensing agreements that discourage technology transfer.

The Indonesian Ministry of Manpower estimate of the number of expatriates working in Indonesia is 28,000. In a country of 160 million this is a miniscule group, among the lowest in any free-market economy. Yet there is constant harping about the need to replace them; there are plans to reduce the number by 15 percent; and also complaints about the slow pace at which they are being replaced. The implication is that they are withholding employment opportunities for 28,000 Indonesians.

Underlying the widespread belief that use of expatriates deprives citizens of high-level jobs and of the experience to prepare themselves for them is the assumption that the number of high-level jobs is fixed. This assumption is entirely mistaken; it denies the possibility of development and growth. On the contrary, there would be more opportunities for Indonesians in managerial and professional positions if there were more expatriates. The initial number of expatriates in a firm is an indicator of the technology transfer potential of the firm's activities. The decline in this number indicates the first step in implementation of technology transfer—their replacement by adequately trained and experienced Indonesians. The main job of expatriates is to implement technology transfer, in large part by training citizens. It does not follow that there should be no restriction on the number of expatriates. Some foreign firms, Japanese in particular, tend to bring in "too many" expatriates and to replace them too slowly, in terms of providing high-level training and experience for nationals. But this is a tradeoff for Japanese willingness to settle for minority ownership. Too low a ceiling or too rapid a rate of replacement would weaken management control and deter investment.

What can host countries do to optimize the MNE contribution to human resource development? First, they must promote continuing investment by MNEs, so that the MNEs will continue training for skills that are scarce and critical for economic development, and so that MNEs will continue to exert demand pressure on local educational and training institutions.

Second, the host countries must encourage foreign investments that introduce new products and new technologies, although not too far ahead of current absorptive capacities. Such investments will help generate new skills and capabilities, rather than simply demand more of the old skills and capabilities.

Third, they must focus investment promotion efforts so that MNEs will be encouraged to concentrate in a limited number of industries. Then the MNEs training contributions and their demand pressures will be large enough to justify the institutionalization of training, and to induce formation or adaptation of responsive educational and training organizations and programs. If investment in new products and processes is too diffuse, scale economies in training do not arise, and turnover of new skills cannot occur.

Fourth, continuing domestic investment is important, to expand the demand for the skills supplied by foreign firms. Without turnover of experienced managerial, professional, and technical employees from foreign firms to domestic organizations, there is little technology transfer, and there is less training by foreign firms.

Finally, as absorptive capacity improves, host countries should encourage licensing of new technology, and the accompanying training and technical assistance, to independent local firms rather than to MNE affiliates. Such licensing assures that there will be incremental human resource development, particularly at senior management and professional levels.

In brief, MNEs should be used as suppliers of development-related skills. But more important, they should be used as stimulants for development of local capabilities and institutions.

References

Baranson, Jack. 1981. *North-South Technology Transfer*. Mt. Airy, Maryland: Lomond Publications, Inc.

Briskey, Ernest J. 1984. Science and Technology: The Shape of Things to Come. *Thai–American Business* (March-April): 20–27.

Goodell, Grace E. 1984. Bugs, Bunds, Banks, and Bottlenecks: Organizational Contradictions in the New Rice Technology. *Economic Development and Cultural Change* (October): 23–41.

Helfgott, Roy B. 1973. Multinational Corporations and Manpower Utilization in Developing Nations. *Journal of Developing Areas* (January): 235–245.

Jamison, Dean. 1975. Radio and Television for Education in Developing Countries. In International Bank for Reconstruction and Development, *Investment in Education: National Strategy Options for Developing Countries*. Working Paper No. 196, pp. 37–49 (February).

Janzen, Daniel H. 1973. Tropical Agrosystems. *Science* (21 December): 1212–19.

Krause, Lawrence. 1982. *U.S. Economic Policy toward the Association of Southeast Asian Nations*. Washington, D.C.: Brookings Institution.

Mears, Leon A. 1984. Rice and Food Self-Sufficiency in Indonesia. *Bulletin of Indonesian Economic Studies* (August): 122–138.

Oshima, Harry. 1984. Issues in Heavy Industry Development in Asia. *Ekonomi Dan Keuangan Indonesia* 32 (1): 31–72 (March).

Stewart, Charles T., Jr., and Jin-Hsia Lee. 1986. Urban Concentration and Sectoral Income Distribution. *Journal of Developing Areas* (forthcoming).

Suriadjaya, B.A. 1983. The Private Sector in the Indonesian Industrialization Process. *The Indonesian Quarterly* (April): 12–21.

Tarrant, James J. 1983. *An Approach to Self-Reliant Rural Environmental Development in the Uplands of West Java: The Case of the Ciamis Program*. Preliminary draft (December).

Thai–American Business. 1986. APCAC Singapore Meeting. (March-April): 12–18.

United Nations. 1984. *Costs and Conditions of Technology Transfer through Transnational Corporations*. ESCAP/UNCTC Joint Unit on Transnational Corporations, Economic and Social Commission for Asia and the Pacific, Bangkok.

National Academy of Sciences. 1973. *U.S. International Firms and R, D & E in Developing Countries*. Washington, D.C.

U.S. National Bureau of Standards. 1977. *Economic Analysis of the National Measurement System*. Barry W. Poulson, ed. A report from the 1972–75 Study of National Measurement System by the NBS Institute for Basic Standards, 1983. NBSIR 75–948, September.

Young, Alwyn J. 1983. *Small Business Development Constraints*. Jakarta: Lembaga Pendidikan Dan Pembinaan Man ajemen (LPPM) (September).

_____. 1984. *Nurturing Under-Represented Groups in a Country's Small Enterprise Development*. Paper presented at the Colombo Plan Seminar, Colombo, 10 July. Jakarta: Lembaga Pendidikan Dan Pembinaan Manajemen (LPPM) (June) 1985.

Yuthavong, Yongyuth, et al. 1985. Key Problems in Science and Technology in Thailand. *Science* 227 1 March 1007–11.

Index

About the Authors

Charles T. Stewart, Jr., has been professor of economics at The George Washington University (from which he received a Ph.D. in 1954) since 1965. Before returning to The George Washington University as research professor in 1963, he taught economics at Utah State University and worked as senior research analyst at The Georgetown University Graduate School and as research economist and director of economic research at the United States Chamber of Commerce. He has written books on low-wage workers, minimum wages (with John Peterson), the supply of medical personnel (with Corazon Siddayao), air pollution and health, and supply-side economics (with Melvin Greenhut), as well as numerous articles on manpower, regional and urban economics, and development.

Yasumitsu Nihei is professor of economics at the Institute of Management and Labour Studies of Keio University, Tokyo, with which he has been associated since 1964. He was visiting research associate at the Institute of Labor and Industrial Relations of the University of Illinois and has been visiting senior lecturer several times at the University of Hong Kong. He received his B.A. and Ph.D. degrees in economics from Keio University and an M.A. in labor and industrial relations from the University of Illinois. He has published three books on labor relations, technology, and employment practices, and several articles on industrialization, technological change, and employment practices.